SpringerBriefs in Applied Sciences and Technology

Series Editor

Andreas Öchsner, Griffith School of Engineering, Griffith University, Southport, QLD, Australia

SpringerBriefs present concise summaries of cutting-edge research and practical applications across a wide spectrum of fields. Featuring compact volumes of 50 to 125 pages, the series covers a range of content from professional to academic.

Typical publications can be:

- A timely report of state-of-the art methods
- An introduction to or a manual for the application of mathematical or computer techniques
- A bridge between new research results, as published in journal articles
- A snapshot of a hot or emerging topic
- An in-depth case study
- A presentation of core concepts that students must understand in order to make independent contributions

SpringerBriefs are characterized by fast, global electronic dissemination, standard publishing contracts, standardized manuscript preparation and formatting guidelines, and expedited production schedules.

On the one hand, **SpringerBriefs in Applied Sciences and Technology** are devoted to the publication of fundamentals and applications within the different classical engineering disciplines as well as in interdisciplinary fields that recently emerged between these areas. On the other hand, as the boundary separating fundamental research and applied technology is more and more dissolving, this series is particularly open to trans-disciplinary topics between fundamental science and engineering.

Indexed by EI-Compendex, SCOPUS and Springerlink.

More information about this series at http://www.springer.com/series/8884

Thomas Hoffmann-Walbeck

Workflow Automation

Basic Concepts of Workflow Automation
in the Graphic Industry

 Springer

Thomas Hoffmann-Walbeck ⓘ
Stuttgart, Baden-Württemberg, Germany

ISSN 2191-530X   ISSN 2191-5318   (electronic)
SpringerBriefs in Applied Sciences and Technology
ISBN 978-3-030-84781-4   ISBN 978-3-030-84782-1   (eBook)
https://doi.org/10.1007/978-3-030-84782-1

This Springer imprint is published by the registered company Springer Nature Switzerland AG
The registered company address is: Gewerbestrasse 11, 6330 Cham, Switzerland

Preface

In this booklet, I am discussing the underlying concepts, models and data formats of *workflow automation* in the graphic industry. I will not present individual or vendor-specific solutions, though they surely have a big impact on the content of this text. That is, I am reviewing industry implementations in general terms – I will not describe workflows in an abstract or mathematical manner.

It is certainly presumptuous to write a book about workflow automation in the printing industry. The field is way too diverse for that. Online printers, transactional printers, commercial printers, gravure printers, in-plant printers, copy shops, wide-format printers and packaging converters, for example, have their own way of producing printed matter. In addition, workflows in the graphic arts industry naturally include additional players, such as ad agencies, designers, publishers, logistics partners, suppliers of consumable, brand owners, and so on. To cover them all would definitely be too much for a small booklet. We are therefore concentrating mainly on workflows in print shops, and to some extent on their direct partners such as print buyers or suppliers of consumables.

Moreover, the need for automation is not the same all over the world but rather depends on labor costs, the local business situation, and quality requirements. That is, I try to describe the state-of-the art concerning the automation of production in the printing industry in general. The actuality in many print shops may differ greatly from this. Moreover, not all products and not all production processes are suitable for automation. Thus, it is not the goal to achieve 100% automation for 100% of products in a print shop. Standard products should be produced automatically, not the very unique one-time jobs.

On the surface, the term *workflow automation* means that everything is cross-linked with everything else, from the creation of the design to the delivered product. Not the automation of individual "production islands" is strived for, but of the overall manufacturing process. Thus, it almost feels like a contradiction if we also analyze small details of parts of the workflow. I therefore would like to ask the reader to always keep the big picture in mind.

In practice, this means that the bottlenecks of the current production flow should be identified first. In print production, workflows are improved step-by-step – it is an ongoing process.

I have given lectures on *workflow automation* in recent years and prepared related lab exercises for students at different universities. These students enrolled in programs on printing technology or something along this line. This brochure is a condensed summary

The author, drawn by Gojko Vladić during lecture

of these activities. Thus, my academic background is still showing through in the content of this publication. I believe, however, that the brochure should also be useful for people working in the graphic arts industry. It helps to understand the relation between different production steps and the intrinsic communication between different components while producing print products.

In Chapter 1, I will introduce the topic of this booklet and define the basic terms that we need in the following chapters.

Chapter 2 is about the scope of workflow automation in the printing industry. In many cases, the term *workflow* refers only to the work steps within production, i.e. in prepress, press and postpress. In this booklet, however, I want to define the term more broadly, which is why I am including areas such as the print buyer's purchase order, materials management, or shipping the finished products to the customer.

In Chapter 3, I present the most common model for print production: the process-resource model. This model is the basis of the Job Definition Format (JDF) that is the topic of Chapter 4. The designation of processes and resources, however, might differ in these two chapters. In Chapter 3, I use terms that are quite common in the industry, while in Chapter 4 I follow the CIP4 organization. For example, in chapter 3, I might denote a resource with paper, while in Chapter 4 I would use the official CIP4 term *media*. CIP4 is a not-for-profit standards association organization specifying data formats for workflow automation (e.g., JDF) in the printing industry. They denote and specify processes and resources so that they are uniformly valid and precise. On the other hand, they are sometimes harder to comprehend intuitively.

Chapter 4 does not only contain an introduction to job ticket formats like JDF, but also other formats such as XJDF and PrintTalk. I explain the basic concepts, not so much the actual coding of these formats. Thus, developers need to study the specifications on the CIP4 website www.CIP4.org (❶) if they want to learn details of coding concerning the data formats.

I do not presume any knowledge of these data formats in order to read this book. On the other hand, I presuppose that the reader is somewhat familiar with the basic procedures in the printing industry as well with the essential concepts of IT technology.

I hope you will enjoy reading this book and that you will find it interesting. Any suggestions, corrections, comments and/or additions would be greatly appreciated. You can contact me at hoffmann@hdm-stuttgart.de.

Thomas Hoffmann-Walbeck

Note

CIP4 stands for *International Cooperation for the Integration of Processes in Prepress, Press and Postpress.*

Acknowledgments

I would like to thank my old friend **Jesko Waniek** who edited the English manuscript with great skill and utmost accuracy. Jesko works as a translator in Marietta, Georgia, USA.

A big thank you also to **Chris Heric,** who gave me great tips on how to embellish the graphics. Chris has been a systems consultant, author, illustrator, and is active in many standards and working groups.

Finally, I would like to thank **Dr. Heiko Angermann** very much, who proofread the manuscript and noted valuable improvements to the content. Heiko is working as a laboratory engineer for printing technology at the University of Applied Sciences in Darmstadt, Germany.

Contents

Abbreviations

AI	Artificial Intelligence
API	Application Programming Interface
B2B	Business-to-Business
B2C	Business-to-Customers
BMP	Bitmap Image File
CFF2	Common File Format, Version 2
CIP3	International Cooperation for Integration of Prepress, Press and Postpress
CIP4	International Cooperation for Integration of Processes in Prepress, Press and Postpress
CRM	Customer Relationship Management
CMS	Color Management System
CSV	Comma Separated Value
CtP	Computer-to-Plate
DB	Data Base
DFE	Digital Front End
DPART	Document Part
DPM	Document Part Metadata
ERP	Enterprise Resource Planning
EXIF	Exchangeable Image File Format
FIFO	First-In, First-Out
FTP	File Transfer Protocol
GPS	Global Positioning System
HTML	Hypertext Markup Language
HTTP	Hypertext Transfer Protocol
HTTPS	Hypertext Transfer Protocol Secure
ICS	Interoperability Conformance Specification
IIOT	Industrial Internet of Things
IOT	Internet of Things
IPTC	International Press Telecommunications Council
IT	Information Technology

JDF	Job Definition Format
ISBN	International Standard Book Number
JSON	JavaScript Object Notation
JMF	Job Messaging Format
MAM	Media Asset Management
MES	Management Execution System
MIS	Management Information System
OEE	Overall Equipment Effectiveness
OCG	Optional Content Group
OPC-UA	Open Platform Communications Unified Architecture
PDF	Portable Document Format
PDF/VT	PDF/Variable Transactional
PDF/X	PDF/Exchange
PDL	Page Description Language
PJTF	Portable Jobticket Format
PPF	Print Production Format
PPI	Pixels Per Inch
RDF	Resource Description Framework
RFQ	Request for Quote
RIP	Raster Image Processor
SAAS	Software As A Service
SFDC	Shop Floor Data Collection
TIFF	Tag(ged) Image File Format
URI	Uniform (Universal) Resource Identifier
URL	Uniform Resource Locator
W3C	World Wide Web Consortium
WMS	Workflow Management System
XJDF	Exchange Job Definition Format
XJMF	Exchange Job Messaging Format
XML	Extensible Markup Language
XMP	Extensible Metadata Platform
XSLT	Extensible Stylesheet Language Transformation

1 Introduction

Abstract *Chapter 1 introduces the subject of the booklet and defines the basic terms that will be used in the following chapters . In particular, we discuss the differences between "workflow automation" and "automation of individual processes" in print production. In addition, workflow automation is briefly considered in the context of "Print 4.0" and "Smart Factory".*

Keywords *Workflow Automation, Process Automation, Print Production, Smart Factory.*

It is generally accepted that driving automation forward is key for the graphic arts industry. It increases quality, shortens production times, decreases material waste, reduces costs, and improves a company's competitiveness in the market.

1.1 Types of Automation

From the bird's eye view, there are two types of automation:

> **Definition**
>
> A **process** is a singular activity with a specific objective that can be planned and executed independently.

A. Automation of individual processes
B. Automation of the workflow

As far as category A is concerned, everyone has a fairly good understanding what a (production) process is. More formally defined, a *process* or *processing step* is a singular activity with a specific objective that can be planned and executed independently. Examples of processes in the graphic arts industry are imposition, printing, or folding. With this definition, neither the make-ready of a press would be a process nor the washing of cylinders/rollers. These activities are not independent but merely sub-tasks of the printing process. They might thus be designated *phases* of a process. On the other hand, *prepress* would not be a process because there are various independent activities involved, such as preflight, imposition, image setting, and so forth.

> **Definition**
>
> A **device** is either a software application or some machine. It initiates the execution of one or several processes. Sometimes, it refers only to the device driver for a machine.

Automation of equipment falls into category A. It refers to the functionality of a machine or some piece of production software. It also includes robots, cobots (for example, in postpress), but most importantly all features of any production equipment that reduce human touchpoints. A classic example of the latter is an automatic register control system on the press. Software modules in *prepress* that execute individual tasks automatically fall under this category as well. Sometimes, the term *component* or *device* is used to cover production machines and software applications that execute processes for print production. Thus, statement A could be renamed

> **Definition**
>
> **Cobots**, or collaborative robots, are designed to share a workspace with humans

© The Author(s), under exclusive license to Springer Nature Switzerland AG 2022
T. Hoffmann-Walbeck, *Workflow Automation*, SpringerBriefs in
Applied Sciences and Technology, https://doi.org/10.1007/978-3-030-84782-1_1

to *component automation* or *device automation*. A machine is then called a *physical component* or *physical device*. Note, however, that some people define a device slightly differently: For them a *device* is only the device driver that interprets external data (such as, data coming from other production components), controls the corresponding machine, and sends data back to other production components. In this booklet, we use the term *device* in both meanings and hope that it will be clear within the respective context.

Note that an increase in the throughput of a device might increase the productivity of the production, but does not necessarily represent an increase in automation. For example, an increase in cylinder rotation of an offset press enhances the productivity of the press (for print jobs with long run-lengths) but would not raise the automation level of the printing process. An inline colormetric measurement and control system in the press, on the other hand, would increase press automation by reducing the need for the printing technician to pull and check printed sheets that often. Similarly, increasing the speed of a software module (for example, by improving the computer hardware) is not a gain in automation, but installing an automatic imposition module would be.

Regarding definition B, a *workflow* is a sequence of defined work steps (processes) in which the activities are normally triggered, controlled and terminated by events. Automation in accordance with category B has two variants. One is purely of an organizational nature, such as determining who is supposed to transport imaged plates to an offset press at what time. The second variant is based on information technology (IT). It encompasses devices sharing electronic data to make sure that processes get the required information/resources they need to be executed automatically, at least in part. Normally, the devices are controlled by one or several central applications called *controllers*, which distribute and collect the data to and from the devices.

Figure 1.1 outlines the situation. In reality, however, it is not necessarily the case that each production area, such as prepress or postpress, has exactly one controller that communicates with all the devices in this area. There can be fewer or more controllers and, in fact, a hierarchy of controllers. Moreover, it is quite common that

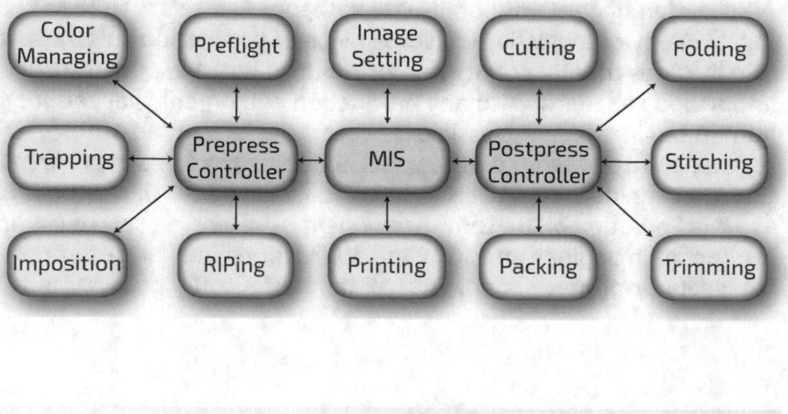

Figure 1.1: Workflow automation example of bidirectional data exchange between controller (gray) and devices (yellow) that execute processes.

Color Managing	Preflight	Image Setting	Cutting	Folding
Trapping	Prepress Controller	MIS	Postpress Controller	Stitching
Imposition	RIPing	Printing	Packing	Trimming

some individual devices are not integrated into the network at all. To have more than one controller in an area is often the case when devices from different manufacturers are deployed, because many device manufacturers require their own controller.

The concept of exchanging data between controllers and devices is also called *process integration*. In paragraph 3.3 we will discuss a production model that portrays this integration. It is based exclusively on processes and resources.

1.2   Resources

Especially during presetting, a device needs data that is generated outside of the device. For example, for each job a printing press needs information about the type, quality, thickness and dimensions of the printing substrate, the primary inks, the printing sequence and so on in order to set up the printing units, the paper feeder and the paper delivery unit of the press. Of course, the press also needs physical assets such as paper, inks and printing forms (plates or cylinders in conventional printing). These are designated *physical resources*. In general, a *resource* is either a physical object or an electronic/conceptual object such as PDF pages or parameter sets. For example, the process of printing might need not only the above-mentioned physical resources but also ink zone presetting profiles, data about the properties of the physical resources, or previews of the printed sheets (the latter for display only or for inline quality control, for example). Please note that in the following I will denote both the physical asset (e.g., paper) and its description (e.g., data about the paper) as a *physical resource*. The valid interpretation should be obvious from the context.

> **Definition**
>
> A **resource** is either a physical or an electronic/conceptual object.

One more prepress example is that RIPing requires information about the requested printing substrate for selecting the correct *process calibration curves.*

> **Definition**
>
> A **process calibration curve** describes tone value changes made inside the RIP to achieve standardized tone value increases in print.

1.3   Interaction of the Workflow Types

In this booklet, the focus is on the IT-related workflow while acknowledging that categories A and B often intersect. For example, sending presetting values for ink zones to a conventional offset press leads to some gain in automation only if the device can handle these parameter values and can preset the ink zone accordingly without an operator stepping in. Actually, the implementation of this particular combination of workflow automation and device automation was one of the first of its kind and dates back to the 1990s.

Workflow automation became more and important in the last decades as the run-length of the average print job decreased and process phases like the setting up of (physical) devices needed

> **Note**
>
> **Process automation** and **workflow automation** are two sides of the same coin.

to be performed more frequently. We mentioned earlier that an automatic make-ready phase of a device normally needs external data – that is, it needs workflow automation. Thus, workflow automation is relevant when a print shop processes many different jobs with short run lengths every day. Thus, it is generally important for commercial offset and digital printing, whereas gravure or security printing, for example, tend to focus on automating individual processes.

You might wonder if workflow automation always needs some device automation to be useful at all. Actually, this is not the case. If a controller comes up with a smart production sequence for different print jobs, changeover times might be reduced tremendously – even if the data is not sent to the device itself but to a computer or mobile device next to it. For example, it shortens the make-ready time of a conventional press if the printing substrate and the order and usage of inks in print modules do not change.

1.4 Print 4.0 / Digital Smart Factory

> **Note**
>
> **Print 4.0** is also known as (Digital) **Smart Factory**. In this context, also the terms Internet of Things (**IoT**) and Industrial IoT (**IIoT**) are used.

Recently, the discussion about workflow automation has also been fueled by the new buzzword *Print 4.0*. The term *Print 4.0* is derived from *Industry 4.0* – an initiative for the entire industry. It is an emergent concept of manufacturing that incorporates the following:

- Mass customizations
- Highly flexible and nearly autonomous production
- Extensive integration of print providers, customers, and business partners
- Linking of production with high-quality services

Technical standards for machine-to-machine communication are the basis for Industry 4.0. JDF, XJDF and PrintTalk play a fundamental role in this for the printing industry. These formats are discussed in sections 4.2 to 4.6.

The vision of Industry 4.0 is that "orders ... book their processing machines and their materials and organize their delivery to the customer" (Spath, 2013). In other words, it is about cross-linking between production components, suppliers, and customers.

Hence, Print 4.0 relates closely to workflow automation. Both involve flexible and automated production. In many cases, however, workflow automation primarily refers to actual production, while Print 4.0 encompasses the entire value chain from order to delivery. Moreover, Print 4.0 stresses the variability of the product being produced. The distinction between the two terms is therefore somewhat blurred. The printing industry currently pursues the following topics in connection with workflow automation and Print 4.0:

1. Integration of production processes and data collection
2. Continuously capturing and analyzing high-volume information (*big data*) from sensors/devices on the production floor, e.g. for predicting events
3. Online job submission and automatic job control
4. Automatic procurement and product shipping
5. Production of personalized products

In Industry 4.0 terms, topic 1 might be characterized as a *cyber-physical system* (CPS). We will deal with the topics 1, 3 and 4 in the next two chapters. In particular, we will present the data formats for the essential data exchange in Chapter 4.

Reference

Spath D. et al (2013) Fraunhofer Verlag, Studie Produktionsarbeit der Zukunft - Industrie 4.0, ISBN: 978-3-8396-0570-7

2 Scope of Workflows

Abstract *Chapter 2 is about the scope of workflow automation in the printing industry. In many cases, the term workflow refers only to the work steps within production, i.e. in prepress, press and postpress. In this chapter, however, the term is defined more broadly, which is why areas such as the print buyer's purchase order, materials management, or shipping the finished products to the customer are included.*

Keywords *Management Information System, Print Job Submission, Web-to-Print, Procurement, Workflow Management System, Printing Technology.*

From the print provider's perspective, there are workflows with

a) internal interfaces/processes which are organized and executed inside the print shop
b) interfaces/processes that relate to external partners

> **Definition**
>
> An **interface** is the point where two processes or software components interact.

Figure 2.1 shows the most important interfaces of a print provider.

The interfaces b) must always connect to internal workflows. That is, workflows can have internal interfaces and processes only, or they can have external ones as well. However, this distinction depends on the range of the considered workflow. For example,

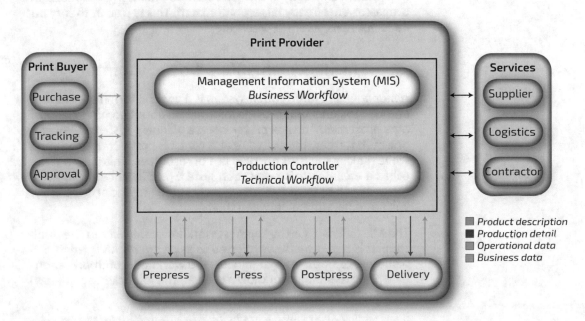

Figure 2.1: Interfaces of a print provider to external business partners and internal production departments

© The Author(s), under exclusive license to Springer Nature Switzerland AG 2022
T. Hoffmann-Walbeck, *Workflow Automation*, SpringerBriefs in
Applied Sciences and Technology, https://doi.org/10.1007/978-3-030-84782-1_2

> **Definition**
>
> A **workflow** denotes the sequence of production and business processes, during which documents, information and task descriptions are passed from one participant to another for action according to a set of procedural rules.

if you look the *plate making* workflow, there are only internal processes at first glance (like imposition, RIPing, imaging), provided it is not done by some external subcontractor. An approval process of the imposed sheets with the customer certainly would count as an external interface. The same holds with a notification that the platesetter has imaged the plates. Thus, it is a matter of defining the range of the workflow if it is internal or external. In the end, if you consider the entire print production as one huge workflow, it always has at least two external interfaces: *purchase* order and *delivery*.

Having said this, I still distinguish between the two. In Section 2.1, let's start with the business workflow in a print shop controlled by a *management information system (MIS)*. I am putting this topic at the beginning because the MIS is often the central communication hub between internal and external workflows. Section 2.2 deals with the communication between the print buyer and the print provider. Next, let's discuss the print shop's material procurement (2.3). Section 2.4 is about the internal print production workflows, which are currently the most dominant area of workflow automation. This is where it all began. As a sub-topic, I will present *workflow management systems* (WMS), i.e., controllers that manage one or several production processes (partly) automatically under a common user interface. Section 2.5 is about the delivery of the finished products. Finally, the chapter ends with other interfaces to external partners to whom a print provider has outsourced parts of its business and production.

Some people claim that the external interfaces allow for major increases in efficiency in the future, while the internal production processes are already largely automated. This is true in theory but does not really apply in practice.

2.1   Management Information Systems (MIS)

A *management information system* is a general term for computer-aided systems that provide information about commercial business processes. An *enterprise resource planning system (ERP)* goes one-step further: the software also controls production processes while making the best possible use of resources. This is sometimes called a *manufacturing execution system (MES)*. Since all three terms are often used interchangeably in the graphic arts industry, I will not differentiate between them.

The MIS/ERP/MES is the topmost workflow controller in the print shop. It is also sometimes referred to as an *agent*. An *agent* is a workflow component that initiates the workflow communication by writing the initial workflow data (i.e., the job ticket, see below) for a print job.

First, I will look at some key MIS features, followed by the interfaces of an MIS (2.1.2). However, these topics are not disjointed but overlap to some extent.

2.1.1   Main Features of an MIS/ERP

The main tasks of an MIS are:

- Price estimation and job management
- Production planning, controlling and scheduling
- Reporting, controlling and visualizing the business status for management decisions

When you launch an MIS, you normally see a list of all current print jobs managed by the MIS. The order management operator can sort this list in different ways. A core function of the job management system is the creation of new print jobs of different types and estimating their prices. For the latter, an MIS must be able to compile the technical processes of the production for each job. The process planning for the production can go quite deep. For example, the MIS might suggest imposition schemes, even for gang forms. Moreover, the delivery date will be registered.

Traditionally, after the MIS order manager has generated a job and the print buyer had agreed on the price, a paper-based *job ticket* is printed, which in turn is then forwarded to the prepress department. The job ticket contains the product description as well as production instructions (not only for prepress). A paper-based job ticket is not state-of-the-art anymore. Nowadays, the MIS sends the job data including the basic production parameters digitally to a prepress *workflow management system* and to other production systems. The *Job Definition Format* (JDF) is widely used for this purpose (see Section 4.3). The digital transfer of job tickets is more efficient and less error-prone. In particular, data does not need to be entered twice anymore.

Usually it is also desired that an MIS displays the production status of all jobs. In order to achieve this, operational data from the production floor must be sent back to the MIS (see Section 2.4.3). This also allows the print provider to control the production costs per job and compare them with the estimated price set in advance. Thus, the profitability can be calculated for each job or job type (post-calculation). Furthermore, it enables production control, job tracking, and the automatic generation of meaningful management reports on things like turnover development, delivery performance, machine utilization, waste analysis, or more abstract values such as *overall equipment effectiveness* (OEE).

Most MIS have optional modules that can be licensed separately, such as a web shop system, an electronic planning board, or warehouse management software. Since these three tasks can be

Definition

In a **gang form** (parts of) several print jobs possibly from different customers are placed on the same sheet.

Definition

A **job ticket** is a document that details the specifics of an order. It can contain a description of the intended print products as well as instructions for their production.

Definition

A **workflow management system (WMS)** is a software system that defines, creates and manages the execution of workflows. It also controls one or more devices which are able to execute processes.

Note

The **overall equipment effectiveness (OEE)** measures the manufacturing performance relative to its full potential, i.e., the manufacturing productivity.

performed with external modules from a different vendor (and be integrated with the MIS), I discuss them not in this MIS section, but rather in Sections 2.2.2, 2.4.4 and 2.4.5, respectively.

2.1.2   Overview of Interfaces

Figure 2.2 shows the main interfaces of an MIS except for the production interface. Note that many of these might not be part of each MIS configuration in a printing plant. They can be either optional modules of an MIS or are not included in the MIS at all. The modules are increasingly merged (e.g., print buyer, web-to-print, and customer relationship management).

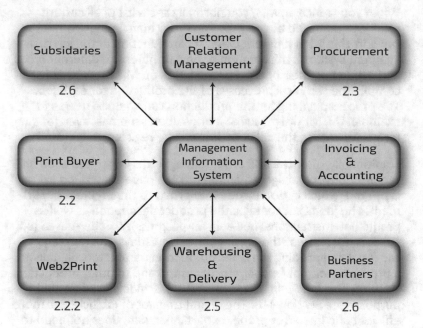

Figure 2.2: (External) interfaces of an MIS

In Figure 2.2, the section numbers are marked for those interfaces that I discuss in the following sections (2.2 through 2.6).

The top arrow between the print buyer and the print provider's MIS in Figure 2.1 represents all job-related communication issues such as quotes, purchase orders or invoicing (see Section 2.2.1). It does not include marketing activities, which are covered under the term *customer relationship management* (CRM). Note that CRM is often understood as the sum of all information and activities around the customer, including order-related correspondence.

The most dominant external communication of an MIS is the interface to the internal production (see Figure 1.1 and Section 2.4). Much data which has been already registered in the MIS must be passed on to the production. Examples include:

- Product specifications such as product type, printing substrate, product parts, binding, colors, number of pages, dimensions of pages, finishing processes (e.g., laminating or varnishing), etc.
- Data about necessary production processes such as printing technology, process details, machines to be used, print runs, etc.
- Organizational issues such as customer information, production schedules, and deliveries.

The description of the processes can be quite general or highly detailed. For example, not only the process names "sheet layout preparation" and "imposition" can be transferred to prepress, but also imposition schemes, precise folding positions, or even ready sheet layouts. The depth of this data exchange is important for evaluating the integration between MIS and production system.

The interface between MIS and production has been subject to lots of debate since the beginning of the 21st century. By now, it is a well-established communication channel. The paper-based job tickets that were common in the past have been more and more replaced by electronic job tickets. A smooth MIS-production interface is therefore essential for workflow automation. The data formats between these two instances have also been standardized, primarily in the form of JDF and XJDF (see Sections 4.3 and 4.4).

Note that the data transfer between the MIS and the production controller and devices is not a one-time matter per print job, i.e. when the MIS hands over the electronic job ticket to the production. Often, the print buyer changes the order afterwards, which requires updating. Sometimes, the order manager can even control some devices directly, like reprinting plates, for example. Moreover, the interface normally is bi-directional. The devices and production controller will a) send back operational data to the MIS (see 2.4.3) and b) send data for updating the job ticket information.

An update might be necessary when some production operator changes some values in the job, or some controller/device adds some data automatically. Normally, the MIS job ticket does not cover all parameters for the production, and those are filled in later with default values, for example.

Please keep in mind that these standardized data formats may be used only for communication between the MIS and the production systems – internally, the data structures may be completely different. In fact, much information will be stored in private databases of the MIS and of the production system, but this architecture still allows the user to connect MIS and production systems from different vendors.

Note

The **product specification** is a description of how the print buyer wants the product to be in the end. It is independent of the necessary production processes.

Note

A **product part** is a component of a product. For example, a softcover book consists of two parts: cover and content..

Note

The bidirectional electronic interface between the MIS and production systems is standard nowadays. The MIS sends product/production data to the production, and the production sends status information back to the MIS.

2.1.3 Current Trends in MIS/ERP

In recent years, many new features have been added to management information systems. I want to highlight just two of them.

In the past, MIS server software and the associated client software were installed on the company's own computers (on premises). In recent years, however, more and more systems have become cloud-based, and clients can use their local browsers to communicate with the remote system. Initially, there were many concerns about data security, but these have diminished.

The cloud can be either public or private, i.e., based on the public internet or on some private intranet. Private clouds are relevant mainly for big companies that strive to interconnect several subsidiaries. Multi-site production and internationalization presumes MIS software that can handle not only different languages and currencies, but that most notably can organize production across several locations with the lowest cost. To accomplish this, some very large companies don't install an off-the-shelf solution from some MIS vendor, but rather build their own system instead. Still, they often use some functionality of a purchased MIS product. That is, they build their own MIS by deploying an API (application programming interface) from some commercially available MIS that they bought from an external vendor.

Another important trend concerning MIS is to allow print buyers to input data into the software directly. Thus, the print provider no longer needs to enter all the data to specify the requested product. This capability makes the following features necessary:

- The MIS can grant the customer access in a secure manner.
- The MIS can estimate production costs fully automatically.
- The print buyer is willing and able to enter the product data into the system.

> **Definition**
>
> **Electronic Data Interchange (EDI)** is a universal term for the electronic transfer of information from one computer system to another.

This procedure surely is only valid for long-term customers of a print provider. Alternatively, *electronic data interchange* (EDI) between the print buyer's ERP and the print provider's MIS, based on a master agreement between the two parties, can be even more effective (see 2.2.1).

2.2 Print Buyer – Print Provider

The information exchange between the print buyer (PB) and print provider (PP) is of a diverse nature. I like to classify them roughly according to the type of print job, such as:

- Unique print jobs
- Template-based print jobs

The first bullet point means that the customer requires a unique print product, not only in terms of content but also concerning the product features like dimension, inks, printing substrate, embellishments, and the like. For that, a special quote must be prepared. Let's discuss this MIS-based approach briefly in Section 2.2.1. Template-based, on the other hand, means that the degree of freedom concerning the artwork and the product might be limited. This case will be the topic of Section 2.2.2. Both kinds might include proofing, portals, artwork submission and notifications, which are covered in Section 2.2.3.

Note

In this text, **artwork**, **print data** and **content data** are used synonymously.

2.2.1 Business Transactions

Figure 2.3 shows some typical data exchanges between print buyer and print provider. Each of these *business objects* or *business data* between the two business partners can be defined by a *PrintTalk* element (see Section 4.4). This protocol can be the basis of an integration between PB and PP. For both the PB and the PP, however, the more critical point will be the extent to which the automation of these business processes can reduce the number of human touch points.

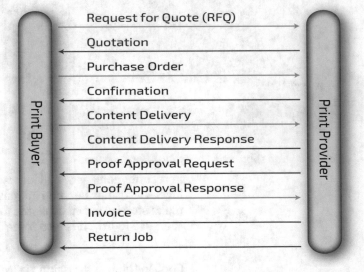

Figure 2.3: Common data interchange between print buyer and print provider

Traditionally, these business objects have been informal and unstructured. The communication channels are telephone, e-mail, postal mail, or even telefax. Direct sales is another high-touch method of placing orders. For the artwork, e-mail attachments or courier/mail services for sending physical media such as CD/DVD or USB sticks are common (Figure 2.4 - ❶). The communication is carried out between PB and PP employees. On the PP side, staff normally re-keys the customer's data into an MIS. These communication channels are not particularly effective. Phone calls and e-mails can lead to misunderstandings and subsequent inquiries. Their potential for automation is low. A change of communication channels should be considered.

To send artwork via an e-mail attachment, files must not be very large. Another option, which has also been used for many years, is to transfer data using the FTP protocol (Figure 2.4 - ❷). For this, the PP must set up an FTP server and the PB can then transfer data to the server using FTP client software.

Note

Business objects or **business data** describe business transactions between print buyer and print provider, such as a request for a quote, a purchase order, or an invoice.

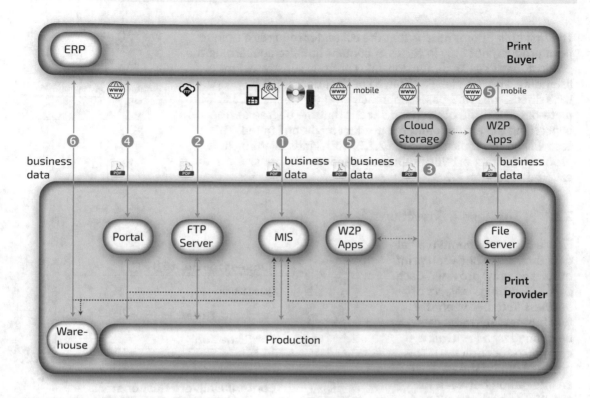

Figure 2.4: Job submission between print buyer and print provider

In recent years, as we all know, the transfer of data via a file hosting services has become increasingly popular. Third-party providers make storage space available for this, which the PB and PP can then use jointly. It is also referred to as *cloud storage* (Figure 2.4 - ❸).

Content data submission via Web-to-Print (W2P) and portals is much more effective. The main advantage is that the uploaded content can be routed to the job data without human interaction. Manual assignments between job data and artwork are no longer necessary. They are not only time-consuming, but also a potential source of errors. In addition, the transfer of the content data is documented and equally transparent for the parties involved. This is particularly important for revisions. Finally, the artwork file can be preflighted automatically, and properties can be checked according the job data (e.g., number of pages or page sizes). The PB will then get immediate notification concerning possible serious errors. Even further processes such as trapping and color management can be automated. In Figure 2.4, arrow ❺ represents this communication path between PB and PP (see next section for more details).

In ❹, the PB uploads an artwork file with a web browser using HTML 5 and some associated programming languages to the portal server. PrintTalk (see Section 4.5) can be used as an underlying protocol between portal/W2P and the PP's MIS. Inside PrintTalk's

ContentDelivery business element, a URL can be defined where the data can be found. The portal/W2P system will send this element to the PP's MIS, which can then respond with a *ContentDeliveryResponse* PrintTalk element.

Some business transactions like *request for quote, quote, purchase order* and o*rder confirmation* are usually carried out before the PB submits the artwork to the PP. With W2P, both the business data and the artwork are sent simultaneously. There is, however, a third option: The print buyer (or some creative agency) submits artwork to the PP and thereby triggers an order implicitly. Usually, a prepress operator or a prepress workflow management system creates the order, and software helps to describe the print job by analyzing the file. For example, the number of pages, page size, colors can be extracted from the PDF (see light green arrow in Figure 2.11). There is still important job data missing, such as the number of requested copies, the printing substrate, or the binding type. The prepress operator or the MIS order manager must get this information from the PB in retrospect. This procedure contradicts the theory but is common practice. It might become even more popular when more product description data is integrated into the PDF (see Section 4.7).

Finally, I want to mention the option of an *electronic data interface (EDI)* between the print buyer's enterprise resource planning (ERP) system and the print provider. This allows a PB to easily retrieve or order products directly via its internal ERP system. This interface is shown graphically in Figure 2.4 - ❻). Such an EDI interface is normally a special solution combining the PB's and PP's systems, i.e., it is customized. It might be connected to the PP's warehouse management software and MIS. However, this type of purchase order is only reasonable for major and long-term relationships between customers and print providers.

2.2.2   Web-to-Print (W2P) and Apps

In the last section, the traditional onboarding methods started with unstructured information from the PB that some MIS order management operator had to re-key. Moreover, the PB or its agency send artwork separately from the job data. Both data sets must then be linked by somebody at the PP, which increases labor costs.

As is well known, Web-to-Print involves creating, ordering, producing and submitting print jobs via the internet. That is, it encompasses e-commerce aspects as well as the creation and submission of the artwork in most cases. Web-to-print automates the submission of print jobs and reduces the number of human touch points for the print provider (e.g. create quotations, process purchase orders).

The main characteristics of a W2P system are:

Definition

W2P is a commercial printing platform for job and artwork submissions using web sites.

- Online pricing, usually online billing (B2C)
- Integration with the customer's ERP (B2B)
- Online customization via pre-defined templates
- Uploading of print data (PDF, text, images)
- Online purchase orders
- Restrictions in product features
- Responsive web design (B2C)

W2P shops and appropriate mobile apps are mainly used for commercial printing, but also – to a lesser extent – for packaging printing, especially for label printing (web-to-pack). Niche markets are deploying this technology as well. The publication of photo books should be mentioned in this context.

In this section, I will go neither into the technical setups of a W2P shop nor into the business opportunities and risks of doing so. In addition, I will not cover the different features and requirements of public B2C and dedicated B2B systems. I will only discuss some fundamental data flow aspects.

One business condition, however, is so fundamental that is has a big impact on the PP's workflow: For online printers, the average value of a purchase order is usually quite small. On the other hand, there are many of these jobs to be processed each day, especially in to B2C environments. With B2B, repeat jobs are common. Both situations imply that order-related activities should be automated as much as possible by minimizing the number of human touchpoints per order. Thus, all the workflow automation options described in this book are particularly important for online printers. Actually, online printers are often pioneers in workflow automation and in some cases the initiators of these new, more efficient production processes. Without workflow automation, W2P is simply not profitable.

The W2P system will inform the PP's MIS and/or workflow management system (see section 2.4.2) about incoming orders. Different data formats may be used, such as PrintTalk with JDF or XJDF, CSV, or private XML (see Chapter 4) to store the desired job data. XJDF might have some advantages over JDF, in particular because a XJDF file can store information about several products, while a JDF file always contains information about a single product only.

To exchange these kinds of data, the W2P server is run by the PP on-site or by some external service provider. If a PP operates its own W2P store, the W2P software is often an optional module of the MIS or prepress production system. In these cases, the interfaces between the modules are obviously internal. For example, the systems could simply access a common database. Of course, there are also PPs that install MIS, production system(s) and W2P software from different vendors.

Definition

With **responsive web design**, the function, design and content of a web page might change in accordance with the screen size and resolution of the computer monitor, tablet, smartphone, etc. being used.

Definition

Business-to-customer (B2C) denotes a business interaction between an end consumer and a company, while **business-to-business (B2B)** refers to a business relationship between two (or more) companies.

Note

A print provider that gets orders (nearly) exclusively via one or several W2P shops is called an **online printer**.

Definition

The file format **CSV** stands for **Comma-Separated Values** and describes the structure of a text file for storing or exchanging spreadsheet data.

Definition

XML stands for **Extensible Markup Language**. It is a very popular way to structure any data in text format. Standard data formats like JDF and XJDF are XML based. Vendor specific XML is also called "private XML".

It is self-evident that MIS, W2P and the production systems must be able to communicate with each other. From the system integration point of view, it is desirable that the vendors deploy standard data formats for this purpose (such as JDF, XJDF, or PrintTalk). This provides the highest level of flexibility in case one of the modules should subsequently be replaced without too much effort. However, even if some of the modules allow only vendor-specific XML, this does not mean that all is lost since XML documents can be easily transformed with help of an XSL transformer. For example, one could convert a private XML document to some standard JDF document or vice versa, assuming that the private XML document contains appropriate information for the JDF.

> **Definition**
>
> **XSL** stands for **Extensible Stylesheet Language**. It is used, among other things, to translate/ transform an XML format into another XML or any text format.

A W2P system is not a monolithic block but usually consists of different modules that are connected to the system via API interfaces. Typically, these modules might be:

- Graphic engine for creating templates and filling customizable templates online
- Rendering engine for generating PDF documents and/or image files
- Preflight engine for checking and making uploaded files production-ready
- DAM for providing digital assets such as images

In addition, of course, there are e-commerce functionalities such as pricing, handling shopping charts, managing previous orders or even a private environment for each customer, data analytics, and interfaces to online payment providers and possibly to shipping companies.

> **Definition**
>
> **DAM** stands for **Digital Asset Management**. It allows the storage, retrieving, organizing and manipulation of digital assets such as images.

2.2.3   Notifications and Approvals

Processes have several different output states – in the simplest case "success" and "failure". Often, some sort of threshold values are set to map the whole range of quality levels up to one of these two extremes.

Most of these validations concerning the processes are handled within the production environment. The checks are carried out either by some operator or some automatic quality assurance mechanism. For example, after imaging (and developing) a print plate for offset printing, either an operator will (occasionally) visually inspect a CtP control strip on the plate or control the tonal values with a hand-held measuring device (mobile plate scanner), or some special image analysis software checks the imaging quality of the printing plates automatically on a conveyor line.

The results of some processes are checked for some kinds of product, but not for others. Trapping in prepress is normally done fully automatically in commercial printing and not inspected by anybody, while in the packaging industry this process will be checked and corrected visually by an operator (because of the higher usage of spot colors and non-standard color separations).

Reviews of processes can also be formally expressed in job tickets such as JDF.

These examples should suffice for internal quality assurance within the production operation. Next, let's take a look at the situation where external people review and approve/reject some interim results – i.e., some resources – during production. The approval stakeholders are the PP, the PB, content providers, and production agencies. On a sidenote, as more and more people work remotely from home, the distinction between internal and external approvals is becoming less significant, at least as far as the underlying technology is concerned.

Traditionally, there are the proofs (i) to (v) of semi-finished products (see below). In this list, I am omitting the design, content and color validations during the creation of the artwork, even though the proofing process is extremely important in this area. The reason for this is that it usually happens independently of the print provider, namely between the print buyer, design agency and photographer (unless the print provider also runs the agency). Moreover, often the focus is more on collaboration. For that, a workspace in the cloud is created, which allows the parties to jointly edit and review online.

i. Checking the PB's artwork for technical production suitability (preflight)
ii. Verification of the print-ready artwork before imposition
iii. Color proof
iv. Imposition proof
v. Sample/mock-up

All these proofs used to be analog on paper or carton, but in many cases they have been gradually replaced by online/virtual proofs. For the latter, a distinction should be made between data exchanges that take place via e-mail or through some special approval software. Approval via e-mail is much more cumbersome and error-prone than, for example, a method that is supported by portal software. It is crucial that the approval process is integrated into an MIS or the production controller or a workflow management system (see Section 2.4.2), because this is the only way to ensure that the right people receive the right proofs on time. Moreover, the software then can warn the PP if an approval request has been ignored or whether a proof has been approved or rejected. The system also tracks revisions. Thus, proofing should be integrated tightly into the workflow software. Let's also note in this context that approval delays are still common, which is why workflow automation in this area is a worthwhile undertaking.

In the following, let's discuss the proofs listed above in more detail.

(i) The visual inspection for technical properties of the PB's artwork is substituted by (or at least complimented with) automatic preflight programs. Nowadays, the print data is not only checked independently from the job order, but the job ticket is used to check compatibility with the artwork (colors, number of pages, etc.). The outcome of this preflight check might lead to an automatic correction, a notification of some operator in the print shop, or a message to the PB (e.g., an e-mail or an entry in a customer portal).

For the sake of completeness, I would like to mention quality inspection systems. They inspect the content of an artwork, for example, the expected legibility of 1D or 2D codes.

(ii) After preflighting, correcting, normalizing, color converting and trapping the PB's data, it is ready for imposition and creating the printing form. However, before the latter happens, sometimes an approval of the print-ready data is requested. In the simplest case, this can be done by sending a PDF file to the customer. This procedure has a crucial catch, however, because the default setting of the customer's PDF-reading software (usually Acrobat) can have a significant impact on the file's appearance. For example, the overprint preview can be switched on or off. Therefore, it is better to send rendered PDF data, i.e. image data, instead. However, the disadvantage of this method is that it involves a lot of data. One solution is to not send a file, but instead to store a preview image in screen resolution on the customer portal (or on some server to which the PB has access) into which the customer can zoom in as deeply as desired with dedicated software. Thus, only parts of the inspected image must be downloaded at higher resolution.

■ Trim Box
▢ Bleed Box
■ Media Box

Figure 2.5: PDF Boxes

Of course, any checks that you can make in a PDF viewer must also be possible with this image-based method, such as the inspection of separations, display of total area ink coverage, measurement of color values, length measurements, showing trim, bleed, and media boxes, and so forth. Another feature of such page inspection software is the comparison of file versions and reshuffling of page positions in reading order.

It is fundamental that the PB can approve, reject and annotate a single page/artwork or an entire file. To accomplish this, the current approval status of each page/artwork must be indicated on the customer portal.

(iii) A *color proof* is a simulation of a print product in the color space of the planned press on the monitor. The goal is to anticipate the future print's color as closely as possible. Since this subject is very extensive, I will not attempt to go into the details of color proofs on inkjet printers or on monitors. That would be a separate topic involving special printing technologies, monitors, lights, calibrations, colorimetry, color profiles, color conversions, and the like, all of which would go far beyond the scope of this book on workflow automation.

(iv) After the imposition process, the (PDF) forms used to be printed out on a plotter. The proof also showed the important dimensions of the trim box, the bleed box, and so on. A prepress operator then checked this *imposition proof* for mistakes in the position of the pages relative to the imposition template, missing bleeds, missing controls marks, etc. Sometimes, this proof was sent to the (internal or external) bookbinder to validate the imposition scheme. Today, only very critical impositions are inspected this way. Mostly, imposition proofs are omitted or reviewed on a monitor after having been rendered. This kind of proof is directed to print production experts only and not suitable for the PB. However, after de-imposing the imposed form – that is, splitting them into pages in the order according to the imposition scheme – the PB might browse through the pages in the reader's order. Sometimes, these are called *signature booklets*. With these, a PB can check easily if the page layout on the sheet is correct.

Definition

An **imposition scheme** (often called **folding scheme**) indicates how pages are placed on a sheet. Only the principle placement of the pages in relation to each other is defined, not the exact position.

(v) In the packaging industry, especially where cardboard/folding boxes are concerned, print buyers often request sample mock-ups. They can be of an analog or digital nature. Analog samples are made with cutting plotters. The digital version would be a 3D simulation of cardboard/folding boxes (see Figure 2.6). An animation allows visualizing and controlling the folding sequence.

For a *Request for Approval (RFQ)* and the actual approvals, the *PrintTalk* protocol can be deployed (see Section 4.5). The PP might send a *Request for Approval* business object, while the PB will send an *Approval Response* object. Both objects need to reference the job ID. To avoid misunderstandings, I would like to clarify (once again) that this of course does not mean that the PB and the PP write, send and read messages in PrintTalk, which is an XML instance. Instead, the software to which the PB has access (such as an e-commerce system or a customer portal) will send PrintTalk business objects when the PB clicks on appropriate icons in

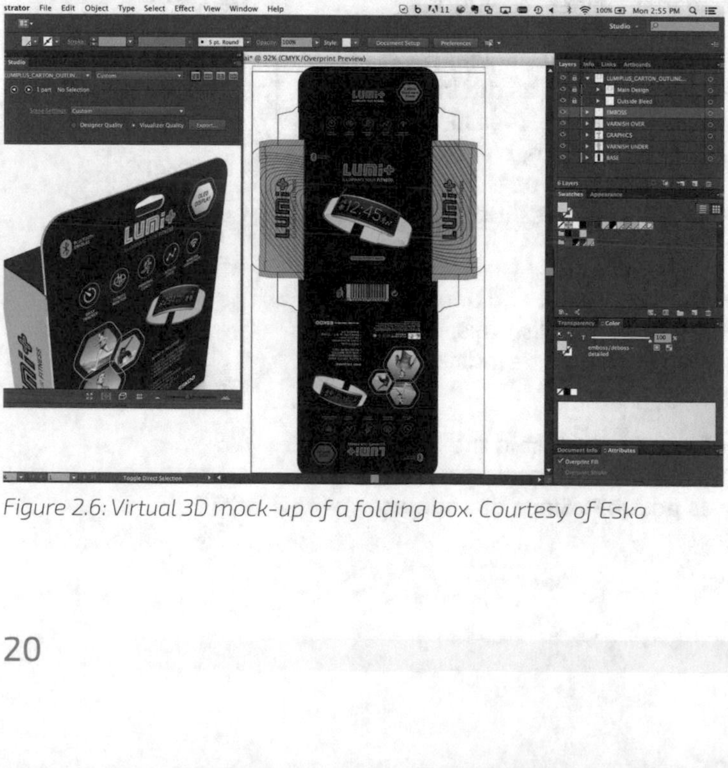

Figure 2.6: Virtual 3D mock-up of a folding box. Courtesy of Esko

the browser. Furthermore, the PP is not notified of the incoming message via SMS or some similar service, but the information is received by the MIS, for example.

During the production of a print job, there are other communications between PP and PB besides the approval procedure. The PP might notify the PB of a preflight report, the achievement of a milestone in the production cycle, or the tracking number of the shipment. Again, these notifications are either displayed on the customer portal or sent via e-mail. As an underlying protocol PrintTalk might be deployed using the business objects *Status Request* and *Order Status Response*.

During production, another message type might be sent from PB to PP, namely an order change. In theory, confirmed orders are immutable, but alas, it is rather common to forward job revisions.

2.3   Material Procurement

Automation in procurement is particularly worthwhile if many different kinds of material are required in small quantities. A web offset print shop that generally uses only a few different printing substrates and plates will have a framework agreement with the material supplier and thus will not need new quotes when material is ordered. The same can apply to a web-to-print shop. In the case of commercial offset printing with short print runs, on the other hand, it makes sense to consider automation in material procurement. In the following, I are assuming such a situation.

Materials management begins with determining whether there is enough of the required material in the print shop's stock. Only if this question is answered in the negative (by the materials management system) must negotiations between the print provider and the supplier of consumables be initiated.

The communication between many print providers and supplier of consumables is still very traditional regarding the purchase of printing substrates, plates, inks, or other materials. Often, some employee in the print shop's purchasing department calls the sales department at the consumables supplier, learns the current prices, and places the order over the phone or later by sending an e-mail/telefax. (Figure 2.7 - ❶) The consumables supplier will then feed the data into his MIS and forward the material to the PP, including the delivery slip. (Figure 2.7 - ❷) The invoice is sent by separate mail. (Figure 2.7 - ❶) The PP needs to manually type the data into the warehouse management system and/or into the MIS.

This procedure is particularly unsatisfactory for the calculation of a quote. Either the PP's employee has a price idea for the printing substrate, or he uses potentially outdated prices in his database (entered during the last order), or he must contact the paper supplier repeatedly. The latter is especially problematic when one considers that many quotations do not lead to a purchase order.

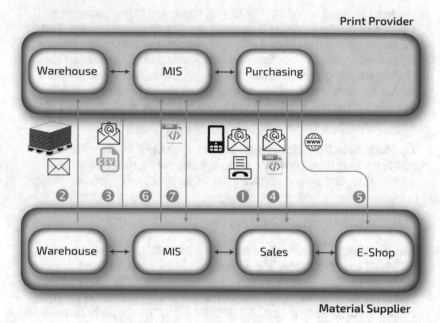

Figure 2.7: Communication between print provider and consumable supplier

There are some very special solutions for streamlining the work-flow:

- The consumables supplier generates a price list from the MIS, for example as a CSV file on a regular basis, and sends it as an e-mail attachment to the PP. The PP imports the list into his MIS in order to make the currently valid prices available for the calculation of quotes. (Figure 2.7 - ❸). The PP can thus update the MIS database for a certain type of consumables (e.g., printing substrates) from this particular supplier.
- The PP sends orders to the supplier as e-mail attachments (in XML or some other text format). This might ease the import of the data into the supplier's MIS. (Figure 2.7 - ❹)
- The consumables supplier invests in an e-commerce system so that the PP can place orders online. This simplifies the processes for the consumables supplier, although the PP still has to enter the request manually. (Figure 2.7 - ❺)
- The PP scans delivery notes and invoices with OCR or auto-matically analyzes e-mails from the supplier in order to sim-plify the data input into his warehouse management system or his MIS.

All of these are very individual solutions for specific situations.

In the future, electronic invoicing might become more popular. The material provider's MIS will generate a PDF invoice, which embeds all relevant data (for example, as an XML record) in a machine-readable format (Figure 2.7 - ⑥). The *ZUGFeRD* specification, for example, makes this possible (Forum elektronische Rechnung Deutschland, 2020).

All these solutions allow only some kind of semi-automatic workflows. The fully automatic solution, on the other hand, is that the PP's management information systems and the consumables supplier exchange XML data directly between their applications via a web service (Figure 2.7 - ⑦). Printing substrates, for example, can thus be ordered fully automatically according to the scheduled printing date in line with the production schedule. Moreover, changes in dates, quantities and articles are also transmitted automatically. Right now, however, such exchanges are not yet standardized, and MIS systems of material vendors have different web service links (if they have one at all). Hence, the PP's MIS must support various formats.

Print providers normally need to communicate with many different consumables suppliers. With each of them, the method of communication may be different, which increases the overall complexity. Possible standardization could be achieved via platforms to which the PP's MIS connects (see Section 2.6).

2.4 Production

The production floor has traditionally been the pioneer in workflow automation. In the 1990s, first attempts were made to automate internal production workflows with the help of a standardized data format (see Section 4.2). At the turn of the millennium, the workflow automation effort was taken much further with the publication of the Job Definition Format (Section 4.3).

Figure 2.4 in Section 2.2.1 shows different paths connecting incoming orders to the production process. For example, there might be a specific method that the MIS uses to send job data to the production (often using the JDF or XJDF data format, see Sections 4.3 and 4.4). W2P orders, however, might be forwarded to the production floor via a different path. This could be caused by different production systems that are deployed for both areas.

There are efforts to unify these channels, for example with the help of an MIS or separate "middleware" software. This module would gather all the orders and print data, assign them automatically, and bundle the data for the following:

- Technical production processes

> **Definition (by W3C)**
>
> A **web service** is a software system designed to support interoperable machine-to-machine interaction over a network.

> **Note**
>
> **Middleware** is a type of computer software that mediates between applications. It can be described as "software glue".

- Orders to external suppliers
- Logistics companies
- Internal storage

This strategy reduces the number of interfaces and software modules within the production. It makes the production more streamlined and cost-effective. It reduces maintenance and increases overall production efficiency. For the latter, it is worth mentioning as an example that creating gang forms becomes more efficient the larger a job pool gets.

Of course, the single connection between MIS/middleware and production shown in Figure 2.8 is an idealized sketch because different printing technologies and print products require different processes and devices. They do not necessarily have to communicate with a universal production controller, which in turn exchanges data with the MIS, but may also be able to communicate directly via the MIS.

Many devices such as presses or postpress machines are individually connected with the workflow network. At times, a specialized controller governs these devices (see Figure 1.1 in Section 1). In prepress, however, applications often cover a few processes, normally under the direction of an operator (see next section).

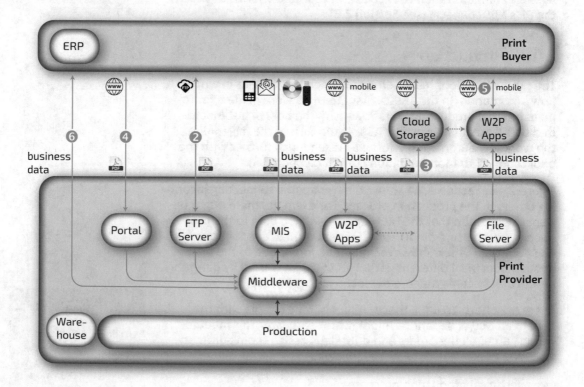

Figure 2.8: Communication between print provider and print buyer

A workflow management system (WMS), executes several different processes. Its software architecture is quite different compared to a stand-alone application. Let's discuss this in 2.4.2. The most networked module in the production environment is the electronic planning board (2.4.4). This software is, in a way, the champion of workflow automation.

2.4.1 Applications for Production

At first glance, it seems that every device and every software application is responsible for a single process. At second glance, however, numerous devices might come to mind that can execute more than one process. A striking example of this is a saddle stitcher, which collects and stitches. These are two different processes. Even a conventional sheet offset press can do more than print these days – it may also be able to varnishing, perforate, die-cut, emboss, and crease. However, it is still true that the main task of such a press is to print. The situation is different with a web offset press, where additional equipment such as a folder and a cutter is standard. Digital printing presses are also frequently supplied with inline finishing modules.

Nevertheless, it is still true for the press and postpress segments that, as a rule, a machine has only one main application, i.e. it executes a single process. In the prepress area, however, things are completely different. Here, too, there are still software programs that execute only a single process, such as preflight or color conversion. In recent years though, the functionality of individual applications has increased significantly. A typical preflight program nowadays can also fix errors of a PDF file, manually or automatically, and convert the file into some standard format (such as PDF/X or PDF/A). Imposition, trapping and color management can also be part of the PDF editor's scope of services.

About 20 years ago, raster image processors together with imposition software have developed into workflow solutions that can cover a wide range of tasks. Such software is often called a *workflow management system (WMS)* or just *workflow system*. Let's discuss them in the following section. The distinction between software application and WMS is somewhat blurred, however. The terms workflow management system or workflow system are now also used for workflow solutions in the press and postpress sector. At the same time, RIP technology has developed into Digital Front Ends (DFE) for digital presses.

2.4.2 Workflow Management System

Individual production processes are often handled not by isolated, stand-alone applications, but rather by a software system. Such a software system controls several modules/devices, and each of them in turn can execute one or more processes. The entirety of all modules and the control software itself is called a *workflow management system* (WMS), the control software is called the *core*, and the individual modules/devices are called *workflow engines*. A WMS is therefore not a monolithic application that fulfils many tasks, but rather a conglomerate of workflow engines that communicate with each other via interfaces and above all with the core of the WMS. The core initiates and controls the execution of the workflow engines (see Figure 2.9).

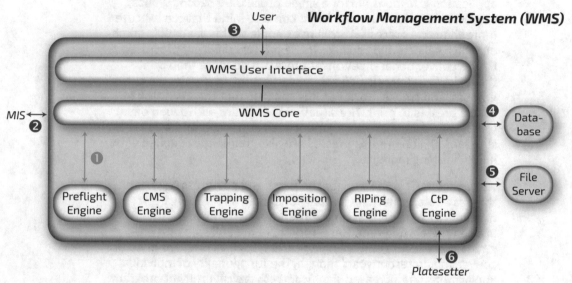

Figure 2.9: Interfaces of a workflow management system in prepress

Three properties characterize a WMS:

- There is a common user interface for all workflow engines.
- A WMS administrator can model the process chains for various production processes within the framework of the WMS.
- A WMS administrator can define default values for the workflow engine parameters.

Generally, the workflow administrator uses a graphical user interface to compile visually different processes into a process chain. Arrows are used to connect the individual icons to specify the sequence and dependencies of the processes. Furthermore, process-specific property tables can provide the processes with instructions and default parameters. Figure 2.10 shows an example of a graphical representation of a workflow. This simple workflow example requires PDF and JDF files for each print job as input. It then sorts the PDFs according to the paper they will be printed on. This information is stored in the JDF file. The workflow management system in this example is *Switch* from *Enfocus*. The special feature of this WMS is that intermediate results between the indi-

> **Note**
>
> **CMS** stands for **color management system**. Color management is the controlled conversion of color between the color representations of various devices.

26

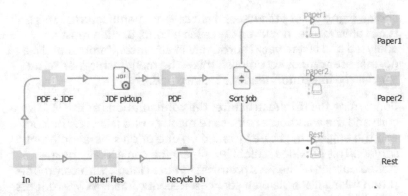

Figure 2.10:
Designing a workflow with
a workflow management
system

vidual processes or phases can be accessed in graphical objects looking like folders, which makes it easy to check whether the processes are doing what they are supposed to.

The workflow core provides the workflow engines with necessary parameter data so that they can execute a process. The core in turn receives the data from the following

- External job tickets, for example issued by an MIS (Figure 2.9 - ❷)
- Result values from preceding workflow engines within the WMS (Figure 2.9 - ❶)
- Default values of the WMS set by a system administrator via the user interface (Figure 2.9 - ❸)
- Values that the WMS operator enters in dialog windows of the WMS (Figure 2.9 - ❸)
- Values extracted from the content data

When you use a WMS, there are at least two user roles: While the administrator sets standard parameters and defines the process chains for various production routes, a WMS operator will change predefined workflows and parameters in a few cases for an individual job only. Standard jobs can thus pass through a WMS completely automatically, at least in prepress. This is especially true for installations at W2P print shops.

Communication within the WMS (Figure 2.9 - ❶) can be handled by:

- Vendor-specific data formats (such as private XML)
- Standardized job tickets, such as JDF or XJDF
- A combination of the two

If standardized job ticket formats are used, a workflow engine is also called a *job ticket processor*. In theory, job ticket processors for a specific process are exchangeable if they are based on the same standard. Thus, during the installation there might be a choice of job ticket processors from several manufactures. This is not always true, however, and such a configuration must be analyzed and tested very thoroughly in advance. Unfortunately, a normal user cannot accomplish this – the manufacturer or some specialized integrator needs to be involved.

Concerning the IT infrastructure, the configurations of the WMS core and the workflow engines are mostly quite flexible. The core and the engines might be installed on one or on several different servers. This provides scalability. An engine can actually be installed outside of the PP's premises "in the cloud", with communication taking place via web services. Recently, many WMS vendors have begun offering their entire system as a cloud-based service (SAAS). There might also be several instances of an engine running simultaneously in order to boost performance.

The communication between the workflow core and the workflow engines is bidirectional, i.e., the engines send status information back to the core.

The data exchange between the WMS and the outside world is normally based on standard formats, such as JDF/XJDF in Figure 2.9 - ❷. This job ticket data would stem from an MIS or some workflow controller. It also might be connected to a portal system in which customers can manage their jobs, upload content data, check monitor proofs, specify correction requests, or give approvals. The information is forwarded to the WMS, and the content data is automatically routed to the print job. Conversely, a WMS can forward the production status to the portal to inform the customer about the production progress.

Figure 2.9 - ❻ represents the interface between the CtP engine of the WMS and the actual CtP device. The engine might forward TIFF-B images to the device, i.e., individual bitmaps for each color separation. Status and error messages might go the opposite way.

Note that a WMS often needs to access external data – either files from the file system (Figure 2.9 - ❹) or records from a database (Figure 2.9 - ❺). For example, the list of all current jobs in the system is usually stored in a database rather than retrieved from the JDF files, which would be possible but quite slow.

In press and postpress, a WMS might only support one type of process/device such as *printing* or *cutting*, respectively. Here the main task of such an WMS would be to translate the standard job ticket format into device specific instructions, balance the workload, and prepare the job offline. The latter allows job-specific settings such as cutting sequences to be defined outside of the actual device on a computer and thus not to block the much more expensive device during this time.

A WMS manages several workflows. From an abstract point of

view, a workflow is managed internally by a system of rules, where a rule consists of events, flow controls, and actions.

Events are triggers like these:

- Job arrived in queue or hot folder or ftp-site or portal
- Approval has been granted
- Sheet is populated with content data
- Plates have been exposed

Conditional branches (if-then queries, loops, etc.) implement flow controls. Conditional branches can be refined by regular expressions.

Actions are activities, such as the execution of a process (by a workflow engine). However, an action could be also just a phase of a process or even just a basic task, such as:

- Extract information from data...
- Convert data into...
- Copy a file to...
- Send an e-mail to...
- Open a dialog box...

Each action receives a status after its completion (e.g., success, warning, error) that can be linked to other actions. That is, each action ends implicitly or explicitly in a flow control. In most cases, the WMS manufacturer predefines the actions, but more experienced users can also write their own in some environments (using programming languages such as *Visual Basics* or *JavaScript*).

Finally, the MIS, a production controller, or a WMS must be able to map print product types to different workflows. A business card, a folding carton box with a lid or a hardcover book will obviously require different workflows and different product parts. In fact, things are even more complicated: There are mandatory, optional and impermissible processing steps for each of these product types. Moreover, the production processes depend not only on the product type, but also on the printing technology. Producing a label on, for example, a flexographic press is different from producing a label on a digital press.

A WMS gathers operational data from its engines, edits and visualizes the data, and passes it to the MIS or to some other controller. The devices can compile different types of data:

- Messages – for example, the current state of an engine such as the current printing speed

> **Definition**
>
> A **regular expression** is a sequence of characters that defines a pattern.

- Reports/protocols – for example, a process summary per job after a process has been executed (how many good/bad sheets were produced in total, etc.)
- Machine control values – for example, the parameter settings of vacuum-powered pickup suckers for the feeder, and delivery or the actual ink zone settings

Messages can be used for progress bars or a status indicator on a dashboard. Protocols are the basis for controlling, reporting, and post calculation. Machine control values are useful for repeat jobs.

2.4.3   Operational Data

Operational data is sent from devices to a controller or from a controller to another controller, e.g., the MIS. There are four main reasons why operational data from production devices is important.

- Scheduling of jobs
- Status indication of production devices
- Job tracking
- Post-calculation and reporting

To schedule jobs, the planning board software needs to be up-to-date concerning the production status at all times (see 2.4.4). In order to be able to react instantly to any production disturbances, the status of each device should be continually known. In 2.2.3, we discussed that in certain configurations the print buyer might be able to check the status of his order, for example via a web portal. To ensure this, the devices must be able to inform the system which jobs they are currently processing and which they have just completed. Finally, to calculate the final cost of a job, data is needed from each device, such as consumed material and elapsed time. A management decision about the profitability of a product type or the efficiency of a device should be based on such data.

As with a WMS, there are two categories of operational data in the current context:

- Reports
- Messages

Reports are records of the result of a process or a process phase after it is executed. A report is static in the sense that it should not be changed later. it is a summary of the activities during the execution. Examples include total material consumption, time stamp at beginning and at the end, and any milestones that were reached. It should contain the final status after completion and can contain the name of the employee who was involved in the process.

The status should cover the device itself as well as the number of product components that were produced, separated into waste and good. It might also include special events during a process or a phase (such as a paper jam in the press) or extra costs that can be charged (for example, due to change of materials or requirements, correcting files, etc.).

In contrast to the reports, messages are issued dynamically. They provide information about the current status of a device, its material consumption, and the production progress. A message may signal the current processing speed or the current status of the process or process phase. In fact, messages can be issued without any process currently being executed on a device at all. They may just inform about the status of the machine (such as idle, breakdown, failure, busy, pause, or repair). The values in the messages can control progress bars and other graphical elements on a production dashboard. They are also used for updating a planning board.

Messages can be issued either spontaneously (e.g., if an error occurred) or at regular intervals (e.g., to provide information about the status of the device or its material consumption). The latter are usually issued without any hand-shaking protocol, sometimes also denoted as "fire and forget".

A device can send messages either on his own or because it is told to do so by a controller. A controller can interactively instruct a device to send short messages on its own at regular intervals. The controller can specify an addressee in this request – either itself or another workflow component such as the planning board. However, it can also request information very specifically from a device. There can be several reasons to do this. For example, the controller might want to get information about the device's capabilities. This can be very general or highly detailed. During a plug-and-play configuration of a workflow, a controller might just want to know which process a device can execute in general. However, the queries can also be much more in depth, such as querying a folder concerning the feasibility of folding paper of a certain size and quality according to a certain folding scheme. The implementation of device capacity protocols will certainly be one of the basic elements of autonomous print production in the context of Print 4.0.

Refining the Figure 2.1, Figure 2.11 is showing a more detailed data exchange between MIS and production systems. A similar figure could be drawn for the interface between a general workflow controller and a WMS.

A device might send the message to the kernel of a workflow management system, to some general workflow controller, or directly to an MIS. A WMS or a controller can forward this information, for example to the MIS or the planning board. It may summarize and/or shorten the content of the messages before doing so.

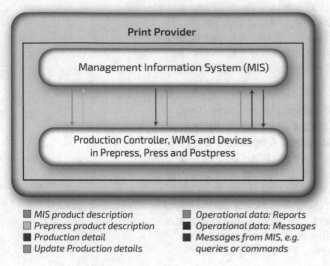

Print Provider

Management Information System (MIS)

Production Controller, WMS and Devices
in Prepress, Press and Postpress

■ MIS product description
☐ Prepress product description
■ Production detail
☐ Update Production details

☐ Operational data: Reports
■ Operational data: Messages
■ Messages from MIS, e.g.
  queries or commands

Figure 2.11: Interface between MIS and production systems

How is operational data collected? There are three options:

- Manually
- Semi-automatically
- Automatically

In modern print shops, it will be a mix of the last two. Manual gathering means in the worst case that each production operator will write on a printed form what he or she is doing. Next, some other person will collect the forms (for example, at the end of a shift) and enter the data into a spreadsheet, which is then used to analyze the production and to generate reports. An already significant improvement is when all operators enter the data into a dialog screen on a (standalone) computer next to the device, and the values are then automatically transmitted to an MIS, for example. Instead of entering data into a dialog screen, barcode scanners are also often used. This replaces the manual transfer of the data from a form to a computer program like Excel. Moreover, the data is gathered much faster, for example at the end of a print job. Finally, the data will be more accurate since it is entered immediately after having finished a process for a certain job and not in bulk at the end of the shift. I like to call this procedure "semi-automatic". In either case, however, the focus is on reporting, messages are not really available.

Real-time messages are only feasible in machine-to-machine communication. The press can signal, for example, its current printing speed or the material that has been consumed so far. Another example of messages generated by a press concerns quality assurance during printing: A camera on the press captures images, and image processing software compares these with the original PDF artwork. Some types of messages, however, an operator may still enter manually, for example the reasons why he or she stopped a press.

Since not all machines can report operational data and not all processes run on a machine or some software application (examples include manual workstations in postpress), there is always the need to capture semi-automatic data within a fully automatic data collection environment. To calculate the actual costs of a product, it is essential to get the relevant data for all production processes.

In many cases, standard JDF and JMF protocols are used for data collection (see Section 4.3), but some manufacturers also use private protocols.

Finally, I would like to stress that not all the operational data of a device is sent back to the MIS. For example, a press might store certain data internally in order to support reprints. Inspection systems and machine sensors generate lots of data that will play an even more important role for the Print 4.0 concept. Analyzing such *big data* allows for the determination of error causes, predictive maintenance, quality enhancements, and generally better machine control. As an added benefit, the knowledge generated this way is not restricted to the direct machine environment. For example, a press could send data to prepress in order to place labels on a sheet in a more print-optimized way next time.

Printing presses may store their data on a central print server. The data exchange protocols used for this differ from the ones we are covering in Chapter 4. The machine-to-machine protocol OPC-UA (OPC Foundation) is used in these industrial applications.

2.4.4 Planning Boards

Often, an MIS also schedules the chronological order of jobs, i.e., the Job queue for each device. The overall objective is to synchronize and determine the production sequence of orders depending on their characteristics and production-related restrictions. A possible variant of this is that the MIS transmits all job data to a separate planning tool (*scheduler*) or to a workflow/production

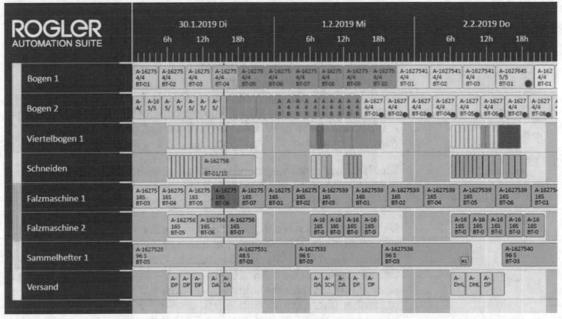

Figure 2.12: A planning board shows the production steps and the status of all current print orders. Courtesy of Rogler AG

controller. The graphical interface of the software displays a planning board that gives the user in the print shop an overall view of the production and planning status in real time. Figure 2.12 shows recent, current and upcoming production steps for orders in a Gantt diagram.

An improved scheduling system can increase the overall productivity of a print shop tremendously. In connection with an offset press, for example, changeover times between jobs (de-inking, blanket wash sequences, changing paper size at the feeder and delivery units, etc.) and the waste of printing substrates can be reduced. Another reason for the importance of scheduling is that the prerequisites for the execution of a production process are triggered in time. For example, it must be ensured that a platesetter images all plates for the printing process in advance. To do this, other processes must have already been executed (preflight, color management, trapping, imposition, and RIPing). Another advantage of a well-organized production sequence is the avoidance of unnecessary breaks between processes. However, required drying times must be taken into account in some cases. Overall, digital planning software delivers a faster production throughput and cost savings, especially in the postpress area, as waiting times and unnecessary material movements to and from the intermediate storage facility can be avoided. Above all, a balanced and smart allocation of orders to the production machines will lead to higher machine utilization.

Ideally, the planning software sends to each device on the shop floor an individual order list that displays the execution sequence. The order list must be synchronized across all devices, of course, to make sure that the outcome of each process is available for the next process according to the schedule.

Another highly appreciated feature of an electronic scheduler is flexibility. Last-minute production changes due to a breakdown of a machine or some reprioritization of a job, for example, no longer have to be initiated manually. The scheduling software will dynamically adjust to the new situation and change the schedule on its own.

An electronic scheduler can be viewed anywhere in the print shop, giving everyone an overview of all planned and executed jobs in real-time. In particular, missed deadlines will be transparent.

Most schedulers have different modes: automatic, semi-automatic, and manual planning. You can manually overwrite planning data that had been generated automatically. Characteristics of production planning can be adjusted, for example by assigning the highest priorities based on deadlines, set-up time optimization, or machine utilization (these constraints might conflict). AI algorithms are also applied to determine the best production sequence. Moreover, you can define shifts and block time intervals for recurring jobs.

> **Note**
>
> **Gantt diagrams** (charts) show the dependency relationships between activities and the current schedule status. They are often used in project management.

A scheduler must be connected to the order administration software (for example, to the MIS/ERP) as well as to the devices in the print shop and the *warehouse management software*. The MIS provides product and production descriptions (JDF or internal data formats) as well as deadlines to the planning software. The latter forwards (JDF) Jobs to the devices in production schedule order. The orders are placed in sequence depending on the equipment and the required consumables. Vice versa, the production devices send status and material consumption information to the scheduler. The schedular must evaluate operational data (both reports and messages – see Section 2.4.3) in order to display production steps/milestones on a dashboard and the current job statuses on a timeline. This makes a scheduler the most prominent hub of workflow integration. In addition, it must of course have precise knowledge of the individual devices and the interdependencies between processes in order to be able to optimize the job sequence and thus achieve a reduction in the transition time between jobs. Obviously, schedulers are also used for partial areas of production, such as for the printing process only.

2.4.5   Warehouse Management

Although warehouse management and fulfillment are often thought of as a single issue, I would like to separate the two and understand fulfillment as the dispatching of goods to the print buyer. We will deal with that in the next section.

A print shop warehouse usually stores more than just finished products, namely printing substrates, consumables such as ink or plates, spare parts, semi-finishes products, end-products, merchandise, and packaging materials.

Probably the most important feature of a warehouse management system involves recording goods in and goods out, for example by using barcode scanners or simple input dialogs on mobile devices. Every material movement should be entered into the system. A proper warehouse management system must also know the storage capacities and be able to autonomously find a suitable place for an incoming item. The storage location of an item is recorded in a database from which it can be retrieved at any time. When items are moved from one place to another within the warehouse, the warehouse software records the movement.

A warehouse management system can also warn of overstocking and understocking. An analogy of a *pipe* illustrates this process (see Figure 2.13.) The entry of an item into the warehouse means that it is filled into a pipe. When the item is taken out of the warehouse, it is taken out of the pipe again. This can happen in parallel, overlapping, and successively. As a rule, it is important that the inventory level in the pipe does not fall below a specific low-water mark, as this would prevent the continuous removal of items on

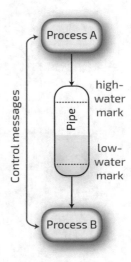

Figure 2.13:
A pipe (buffer) with low-water and high-water marks

the other end. If the article is a semi-finished product or the end product, falling below the low-water mark might indicate that the requested number of units is not available. Of course, the level of the pipe should also not exceed a certain high-water mark, because then the storage capacity may no longer suffice and the warehouse may be filled with excess items, which ties up capital, signals overproduction, and generally increases storage costs. Each article has its own pipe and own high-water and low-water marks that must be set individually.

The water marks should be modified dynamically, for example in accordance with the production forecast if the article is some raw material. If it is a semi-finished product or the end-product, the requested number of copies of the print order(s) or some forecast relating to this should control these marks. In both cases, the MIS needs to feed the data to the warehouse management software.

If the level of an item falls below the low-water mark or exceeds the high-water mark, appropriate departments must be notified. For example, if the level of a print substrate falls below the low-water mark (the minimum stock level), the material procurement department should be notified. It is conceivable, of course, that this will also trigger and automatic order to material supplier.

The MIS should book raw material as early as possible, for example when a quotation becomes a print order. During production, the MIS should also ensure that the material is retrieved from the warehouse in a timely manner. Finally, the MIS should trigger the transfer of the final products to the fulfillment department. This is particularly true if a consignment store is run.

The CIP4 organization has defined the mechanism of dynamic pipe control in its specifications irrespective of any warehouse storage. If an output resource of process A is the input resource of process B (see Figure 2.13), the JMF/XJDF pipe control mechanism might be deployed. However, this does not necessarily mean that the resource is brought into the warehouse and then retrieved again. It could be buffered on a conveyor belt or simply somewhere on the shop floor.

A warehouse management system can be very powerful in connection with an automatic transport system and automatic paper feeding of a press. But even without it, the internal transportation of material on shop floor can be simplified significantly, for example by installing a barcode system for pallets. The software creates a pallet label for each pallet with a unique barcode that is scanned at each workstation. In this way, the system knows the locations of all pallets and can create transport orders independently. To create such orders, it must be able to communicate with the planning board (MIS, production controller) since this the only entity that knows the order of the production steps.

Similar to the planning board, the warehouse management software can be an optional module of an MIS or stand-alone software. I would like to mention in this context that warehouse management software is often browser-based or app-based nowadays, supports mobile devices, and can run on-premises or as a cloud-based solution. For the latter, the communication between client and server should always be encrypted.

2.5   Fulfillment

At first glance, fulfillment may seem simple, but alas, it is not. Let's look at some of the complexities involved in executing a single job:

- Multiple products, shipped on palette(s) or in parcels
- Multiple delivery addresses
- More than one shipment
- Different product quantities to various addresses
- Multiple departure times for the shipments
- Different packaging requirements for the shipments (local, international)
- Various procedures for the shipments, such as custom declarations for international shipments
- Different carriers for the shipments
- Addition of pre-produced or externally produced products from the warehouse

The range of automation in delivering and transport is huge. For example, a prepress operator could simply ask a local bike courier

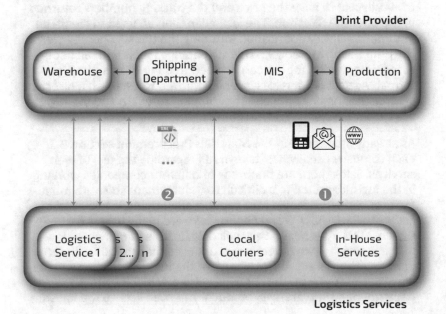

Figure 2.14:
Communication
between print provider
and logistics services

service to deliver printed proofs to the customer (Figure 2.14-❶). On the other hand, there are highly automated fulfillment solutions.

A very traditional way of sending printed products is that an MIS – or even just some text processing software – prints a mailing list, address labels and palette labels on a (label) printer. Especially if a single printed matter is shipped to numerous customers, it is quite convenient to use a list of addressees that the PB has provided in a table. The employees in the shipping department pick the individual items, pack them, label them, and tick them off on the list.

Shipping software, which is often a module of the MIS, can operate more efficiently. For example, the mailing list is stored digitally, and filling the list is done with the help of a barcode scanner. As soon as all required items are registered, the delivery bill is automatically created, printed, and forwarded to the billing address. As soon as the loading of a truck is completed, the software prints the loading list automatically.

Packing and palletizing may be done either manually by an operator, semi-automatically, or fully automatically by robots. Robots are mainly deployed if the print shops processes jobs with high print runs. Of course, the packing of personalized bulk deliveries is completely different and consists essentially of enveloping and addressing.

Special attention should be paid to the interface between the PP's shipping software and the logistics provider that transports the goods.

The aim is to send the shipping data to the logistics company electronically and to have the price and the tracking numbers returned in the opposite direction. The latter can then be used to create an automatic delivery notification including the tracking number for the print buyer. Moreover, accurate package labels with all necessary information for the carrier need to be printed in the right format. Again, web services can be deployed to establish a direct communication link between the shipping software and the logistics carrier's system (Figure 2.14 - ❷).

As already mentioned in the Materials Procurement section 2.3, each company usually has its own API regarding the use of web services. Since there are hundreds of different companies working in the logistics sector, it is difficult for a fulfillment software manufacturer or an MIS provider to support all of them. While everyone knows USPS, DHL, UPS and FedEx, there are many others (consigner, 2021). As a result, they will only be able to manage a few of them, which is why there are companies that specialize in reconciling the different API interfaces of the logistics carriers with a universal interface for the shipping software/MIS (Figure 2.15 - ❶). A distribution software/MIS producer can then license the logistics

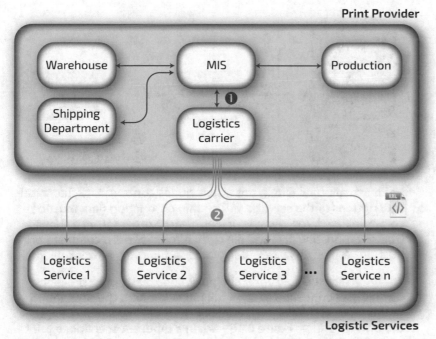

Figure 2.15: MIS/Shipping software with 3rd party module for data exchange using Web services

carrier software as some kind of middleware between it and the actual logistics carriers.

The interface between the shipping software/MIS and a specific logistics carrier module can be very simple. For example, the communication channel might consist of just two hot folders (one for each direction) in which text files are transferred. The text will be structured, of course. For example, CSV, XML or JSON files are used. If the MIS does not incorporate the shipping software and an external software is deployed instead, a similar interface is needed between the two. The data exchange ❷ in Figure 2.15 is often based on web services.

The logistics carrier module may or may not provide an automatic calculation to determine which carrier is the most reasonable company for sending a particular shipment. However, individual rules can be set up for choosing the carrier. Particularly if the shipments involves multiple countries, it becomes time-consuming and error-prone to work out the optimal delivery method for each shipment yourself.

> **Definition**
>
> The text-based, structured data format **JavaScript Object Notation (JSON)** is similar to XML.

2.6 Subsidiaries, Print Brokers, Subcontractors

Managing orders from subsidiaries of a print shop, forwarding orders from print shop to print shop, or communicating with subcontractors and print brokers have one thing in common: There is a need for forwarding job tickets or parts of them from one stake-

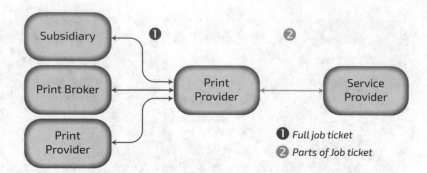

Figure 2.16
Job ticket exchange be-
tween print provider and
partners

holder to another. Often, this also includes content data. Technical descriptions of the services and certain production data must be communicated, for example, if a print shop orders from a subcontractor embossing dies and die cutting tools or even complete production steps such as exposing plates, laminating, or bookbinding.

Theoretically, the concept is simple. A subsidiary, print broker or some other print shop sends a full job ticket with the description of the product details (e.g., PrintTalk with XJDF or simply CSV) to the print provider (Figure 2.16 - ❶). In a different scenario, a print provider sends parts of a job ticket containing product information as well as necessary production details to a service provider (Figure 2.16 - ❷). The PP in the first case respectively the service provider in the second case will ideally import the data automatically into his own system, produce the product or some component of the product or a production tool, and then might send (parts of) the updated data back.

Please note that there might be two different independent software systems involved, namely the MIS of the sender and the MIS of the receiver. Even if they are using the same software, it might not be trivial to connect the two. Only if a subsidiary and the head office both make use of a single instance of an MIS in the cloud can they share data easily by using a common database as long as MIS has been programmed to support multiple print shops.

In most cases, it not too hard to implement sending/receiving entire job tickets between A and B automatically in a secure way. This is especially true if the complete job ticket data exchange between branch offices or between a print broker and a print provider involves large companies, which can set up these connections. We already mentioned in 2.1.3 that these big (online) print shops develop their own MIS in some circumstances using the functionality of some commercially available MIS as a base (via API).

Things become much more difficult if a PP must extract some parts of a job ticket for a service provider and forward the data to him automatically. It is even more complex for the service provider to automatically import those job ticket parts from different print providers. Therefore, such data exchanges are still mostly done via e-mail, phone, ftp and the like in the graphic arts industry. In some cases, there might be customized solutions between two partners, for example via self-written scripts.

Since the job tickets are forwarded electronically, it is important to encrypt the transmission, whereas the data is often physically stored in a virtual private cloud (VPC).

> **Definition (W3C)**
>
> A **virtual private cloud** (**VPC**) is a private cloud within a public cloud environment.

Thus, the industry uses human-readable data (e-mails, etc.) and machine-readable data (CSV, XML, JSON, JDF, XJDF, PrintTalk) for the job ticket transmissions between stakeholders. However, even machine-readable data can have different levels of automation. For example, a CSV file could be assembled manually. Even if the exchange file is created automatically, for example by an MIS, the question remains whether

a) the data is automatically forwarded to the recipient (and not through an e-mail attachment) and
b) whether the recipient can automatically import the data back into his system.

Lately, online platforms have emerged which connect participants such as print buyers, print providers, and partners, e.g., suppliers of consumables, machine manufacturers, and service providers. To achieve this, any software connected to the platform can store product and production data on it. The platform will pass this data on to suitable partners, who will then receive a notification of it.

References

Forum elektronische Rechnung Deutschland (FeRD) (2020), ZUG-FeRD 2.1 1 Specification. Available at: https://www.ferd-net.de/standards/zugferd-2.1.1/index.html (Accessed: 15 June 2021).

OPC Foundation, https://opcfoundation.org/about/what-is-opc/ (Accessed: 15 June 2021).

3 Workflow Models

Abstract *Chapter 3 is about models that are used in print production. First, flowcharts are briefly discussed. After that, the most
common model for print production is explained in detail: the process-resource model. This model is the basis of the Job Definition
Format (JDF) that is the topic of Chapter 4.*

Keywords *Production Model, Process-Resource Model, Flowchart,
Activity List.*

Since automation eliminates the ability to make ad-hoc decisions
during the production of a print product, it goes without saying
that it requires rules on how a type of product is made. While
rules are always needed, even without production automation,
the difference between non-automated and automated production is that for the latter the rules must be formally defined. It is
not enough for the employees to have these rules in their heads.
Moreover, these rules should not always be created anew, though
perhaps differently for different product types. The rules should be
designed according to a uniform scheme – in other words, based
on a model.

Of course, this does not imply that the manufacturing process of
a print product must always follow these rules no matter what, or
even that a production that has been started cannot be changed
or stopped. And since order changes are quite common even after
the production has started, or for technical reasons, it must also be
possible to change the production rules for a specific order. In addition, it must be possible to intervene manually if individual process parameters do not correspond to the product's specifications.
Furthermore, automation may be questionable in an environment
with more one-off print jobs than standard print jobs.

As usual, there are different options for modeling a production. Let
us look at three of them.

3.1 Activity List, Process Chart

Perhaps the first thing that comes to everyone's mind is to simply
list the required production processes, i.e., to create an *activity list*
or to-do list. That is fairly easy and great for brainstorming, but not
so good for a structured model. Obviously, there is no order specifying which process should be executed first, which one is next,
and so on.

An ordered list or a process chart can do the trick. These are actually quite common and sufficient for the first draft of a workflow
description. A process chart is even better because it also allows

© The Author(s), under exclusive license to Springer Nature Switzerland AG 2022
T. Hoffmann-Walbeck, *Workflow Automation*, SpringerBriefs in
Applied Sciences and Technology, https://doi.org/10.1007/978-3-030-84782-1_3

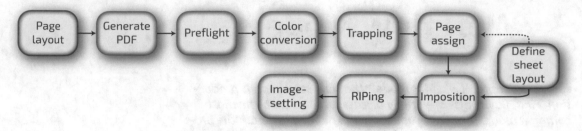

Figure 3.1: Process chart for offset prepress

branching and multiple starting points. Figure 3.1 shows such a chart for a typical process sequence in the prepress phase of offset printing. In this figure, I omitted the creation of layout elements such as text, images, and graphics. There are two starting points: "Page layout" and "Define sheet layout".

Process charts are still too vague, however. Neither the prerequisites (inputs) nor the outcome (outputs) of a process are defined. For example, what is the outcome of the "page assign" process? Is it

- a list of PDF files and page numbers,
- a link to a PDL file with pages in the reading order,
- or an assignment of pages to the position of the sheet layout templates?

What kind of information is needed for defining a sheet layout? What is it that the imposition process and the page assignment process actually do? What kind of data is needed for image setting? Obviously, an automated production system requires unambiguously defined processes as well their inputs and outputs.

The chart shows only an "ideal" workflow. However, every process has at least two status outcomes: success or failure. Figure 3.1 contains neither proofs nor reviews, but they could easily be added.

3.2   Flowchart

Flowcharts are well-known and widely applied. They are composed of symbols and are used commonly in data processing. They serve primarily the visualization of a program designs during software development. However, they are also used in more general models of business and production processes. They became a DIN standard way back in 1966 (DIN 6001, 1966) and were adopted as an ISO standard in 1985 (ISO 1985).

Figure 3.2 shows an example. The blue graphical elements symbolize the start and the end. The green diamond-shaped symbol represents a *decision*. The question allows only two mutually exclusive answers, *yes* or *no* (alternatively *true* or *false*). The red element describes a *predefined process*, which is ideally again described by a separate flowchart. In addition to the three graphical symbols presented, there are of course a few more. A circle represents a *connector* (see Figure 3.3), a simple rectangle indicates a process or operation, and so forth. Please note that only the shapes of the elements are specified in the standard, not the colors.

Figure 3.2 demonstrates a typical communication exchange between PP and PB, in case the PP generates a proof.

In contrast to the activity diagram of the last section, in flowcharts *decisions* are part of the concept. In this respect,

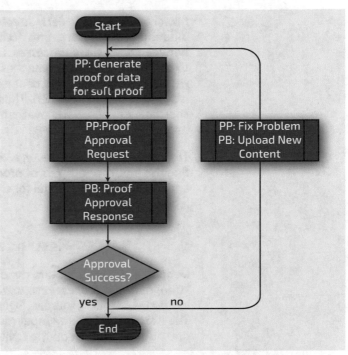

Figure 3.2: Proof cycle between print provider (PP) and print buyer (PB). Source: PrintTalk 1.5, Figure 3

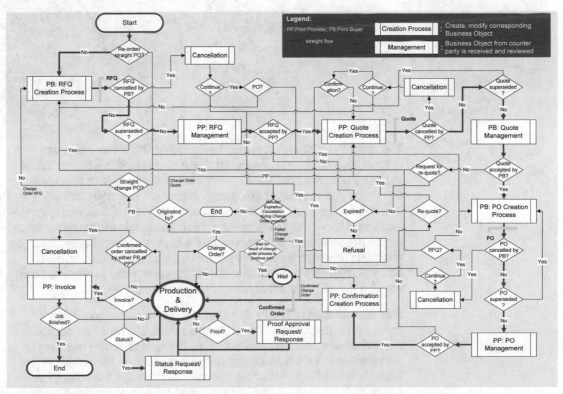

Figure 3.3: PrintTalk workflow diagram. Source: PrintTalk 1.5, Figure 6 , © 2000–2020 CIP4

flowcharts are particularly suitable for modeling flows that take different process states into account. Figure 3.3 displays *decisions* between *predefined processes*. Even though it might be important to consider all possible branches in a workflow, things can quickly get very complicated and confusing if an entire production model is supposed to be specified. Thus, flowcharts might be better for specific parts of a workflow than for the interaction of large numbers of processes.

The decisive factor, however, is that flowcharts usually do not define the inputs or outputs of processes. This leads us to the last and most important model for the graphic arts industry in the following section.

3.3   Process-Resource Model

Defining preconditions and the outcome of processes is the strength of the *process-resource model*. Here, every process has input and output resources. Figure 3.4 shows an abstract example. The process 1 has three input resources on the left-hand side and one output resource on the right-hand side.

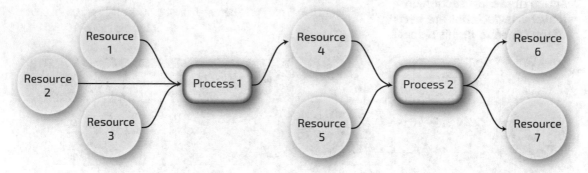

Figure 3.4:
Abstract example of a
process-resource model

The yellow rectangles represent processes; the blue circles are resources. We already discussed these two terms in Section 1.1. A slightly inaccurate but easy take-home message is that processes are verbs and resources are things. A process normally has several input resources, i.e. resources that are needed for the process to execute. Moreover, the process generates one or more resources, which are called output resources. Processes and resources always alternate. A process can only generate a resource, not another process. The similar statement holds for resources: A resource can only be input of a process, not of another resource.

Please note that resource 4 in Figure 3.4 is simultaneously an input and output resource, while the others are input or output resources only. Resources 4 is the most important one because it determines in which order the processes must be executed. In this text, I will call such resources "transitional". In the following, I will concentrate mainly on transitional resources in order to simplify our charts. That is, I will omit quite a few input resources in the diagrams.

The process-resource model is the basis for workflow automation in the printing industry. The basic idea is this: A process may start automatically if all input resources are available.

Digression: Let's make pancakes

Please feel free to skip this digression. It is not essential for the understanding of the following but only supposed to explain the model with the help of a somewhat odd example: making pancake. It proves, however, that a process-resource model is useful for all kinds of projects.

Figure 3.5 shows four processes: *Mix*, *Fry*, *Separate,* and *Beat*. Each process has several input resources. The mixing process, for examples, has:

- Consumable resources, i.e. ingredients, such as sugar and flour
- Tools, such as a bowl and a stirrer
- Operators, such as the chef

Four resources are transitional. They are marked with a red border. The resource *Pancake* is the final product and hence an output resource only.

This example of making pancakes is not complete. Some details are missing. In particular, I did not...

- ...define details of the end product, i.e. the number and diameter of the pancakes.
- ...specify any quantities, for example, the number of eggs.
- ...describe the tools, for example, the diameter of the bowl.
- ...specify the kind of class a resource belongs to; for example, the *chef* is an operator and not an ingredient.
- ...specify the production times. Obviously, a production end for a process needs to be observed in order to meet the delivery date of the final product. On the other hand, a process should not start too early, because of
  (i) last-minute changes that might be required and
  (ii) the output resource of the process might become unusable or damaged if it is stored for too long. Moreover, storing items increases costs.
- ...identify details of the processes, for example, how long they should execute (such as mixing or beating).

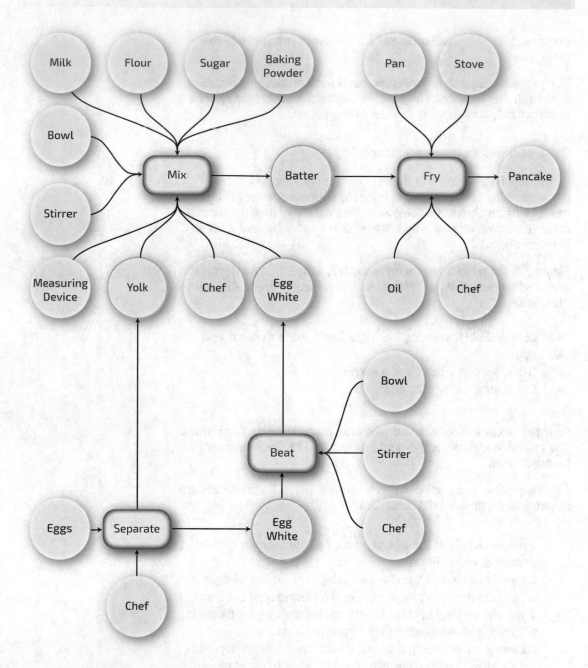

Figure 3.5: Process-resource model for making pancakes

The last point means that usually each process needs one more *parameter resources* as input, which describe further details.

In the context of automation, we can replace the chef by some robots or devices. That is, one robot for each process. To ensure such a production, each process and each resource would have to be specified much more carefully. Moreover, the goal would be to automate cooking in general, not just the production of pancakes. Thus, we need to specify all processes and resources for cooking.

Moreover, robots need to understand this be able to execute the processes automatically. They should be able to communicate with each other and/or a central controller. The specification of the processes and resources should be based on an open standard so that the bakery can buy robots/devices from different vendors – the mixing robot from vendor X, the frying robot from vendor Y, etc.

Let us assume that there is a specification and that there are vendors who manufacture such robots. We still need someone who prepares and integrates them into a production line, plans and reacts to new requirements (for example, when a new recipe is introduced), and intervenes if something goes wrong. This is a task for a workflow manager in a fully automated baking plant. In addition, a workflow manager must think through the entire process chain from start to finish in advance. Manufacturers mainly consider the production segment of their own workflow solutions only.

End of Digression

Normally, a process relates to many input resources and only to one or two output resources. Figure 3.6 shows the model of conventional printing using plates such as sheet-fed offset. This example comes very close to the more general model of CIP4. However, as already mentioned in the preface, I will not use the abstract and general resource names of CIP4 in Chapter 3 but rather the terms commonly used in the graphic arts industry.

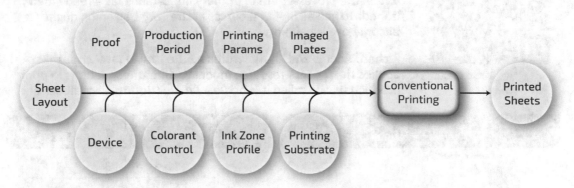

Figure 3.6: Process-resource model for printing process

Still, the resources in Figure 3.6 might need some further explanation:

- Imaged Plates: In lieu of plates there could be cylinders or sleeves, above all for gravure or flexographic printing.
- Printing Params: Device setup information – for instance if the fountain solution modules in an offset press or the heating modules should be turned on or off.

- Production Period: Time interval in which the printing process should be executed.
- Proof: Page or imposition proof produced by the prepress department.
- Sheet Layout: Types and positions of control elements on the sheets – for example, for registration or color control.
- Device: ID of the press and its capabilities – for example, its maximal run speed.
- Colorant Control: Order of the inks.
- Ink Zone Profile: Information about the amount of ink that is needed along the printing cylinder for offset presses with traditional inking units (no anilox rollers) for each separation.
- Printing Substrate: Details about the paper or foil, such as thickness, type, and size.

Figure 3.6 does not show all possible input resources for the process. For example, in flexographic printing the mounting tape should be specified as well. Input resources are mostly non-mandatory. Ink zone profiles, for example, make no sense for an anilox offset press.

The last figure shows many input resources for a single process. For the initial planning of the production workflow of a specific product type, the processes and the transitional resources between the processes are of primary importance, as already mentioned. To add all input resources at the very beginning might become overwhelming.

Video about Workflow Puzzle

Richard Adams and I have written a little tool in JavaScript that applies the process-resource model to different situations in print production. We called it *Workflow Puzzle* (Hoffmann-Walbeck and

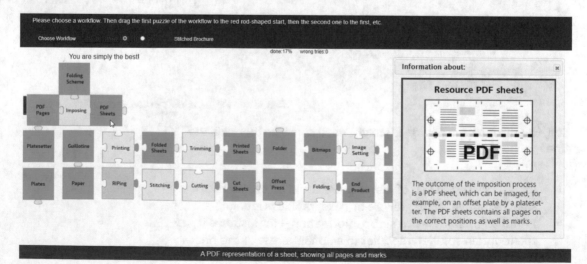

Figure 3.7: Interactive tool for creating a process-resource model

Adams, 2021a). Please try it out yourself. You will find it at (Hoffmann-Walbeck and Adams, 2021b)[1].

In this script, there are different workflows you can choose from, such as Stitched Brochure, Deck of Cards, RIPing, LAMS Flexographic Plates, etc. Figure 3.7 shows the workflow of a stitched brochure. As before, processes are yellow, resources are blue. Holes and bulges define the input and output relations between neighboring cards.

The idea is to drag these cards into the right order. The little red bar is the starting line. The workflow runs from left to right. If you pick the correct card, move it close to the predecessor card and release the button, it will snap into its position, and you will hear a sound. There is also text-based feedback. If you do not know one of these processes or resources, you can get information about it by double-clicking on the card. An information window such as the one on the right-hand side in Figure 3.7 will pop up. In addition, if you hover over a card, a short explanation appears at the bottom of the browser window. In the screenshot, there are four cards already placed on the correct position; all others are random. The software counts the wrong attempts. In my opinion, the best players are those who have the fewest false tries. Unlike with other online games, speed is not an issue.

Figure 3.8: Processes and resources inside the process group RIPing

In Figure 3.7, the *RIPing* card has a slightly different color compared to the process cards because, strictly speaking, RIPing is not a process but a process group. It includes several processes such as *Interpreting*, *Rendering* and *Screening*. Figure 3.8 shows the RIPing workflow in *Workflow Puzzle*.

By the way, at (Hoffmann-Walbeck and Adams, 2020a)[1] you will find a similar tool called *Workflow Arrows* (Hoffmann-Walbeck and Adams, 2020b) that allows you to define processes, process groups and resources yourself and connect them freely (see Figure 3.9).

In the first attempt, there is no need to distinguish that carefully between processes and process groups. However, when the model is refined, we often need to replace the process group with individual processes. This is true not only in graphical modeling on paper, but also in a real production environment. An MIS may not be able

---

1   This script is merely a prototype and not a professional program. We tested it only on a Windows PC and on a Macintosh. Especially on touchscreen devices it will not work well.

Video about Workflow Arrows

Definition

Interpreting is the first process inside a RIP. PDL data like PDF is read, parsed, and simplified for the next process. For example, Bézier curves are iteratively approximated by polygons, images are decompressed etc.

Rendering converts graphical elements, which might include vector data, into contone images (byte maps).

Screening converts contone images to monochrome images (bitmaps).

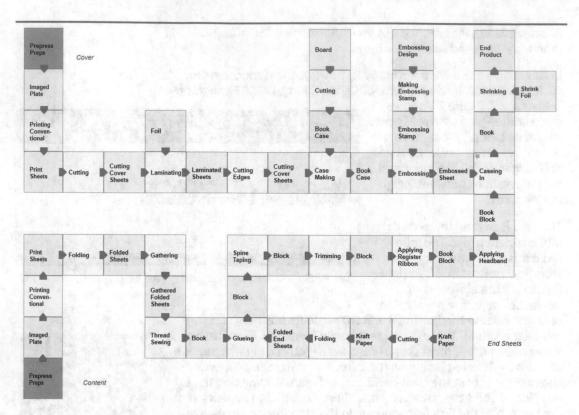

Figure 3.9: Process-resource model for book making

to specify processes individually but might only create process groups such as RIPing. A production controller for prepress that knows the important details will subsequently fill such a process group with individual processes. An MIS might even issue a very general process group such as *Plate Making* or *Prepress Preparation*.

Figure 3.10 provides a more complete picture of processes and transitional resources for a prepress workflow in commercial printing. It does not show the artwork creation processes but rather starts with PDF pages which a PB often sends to a PP.

Unfortunately, this visualization of a process-resource model can easily lead to a misunderstanding. The arrows do not imply that files are copied into the "input area" (such as a hot folder) of the next process. Often, the first process will store the output resource on a server and merely notify the next process where the data is located. In fact, passing information from one process directly to the next is also likely to be an exception. Usually, the process will only inform the MIS or the production controller where it has stored the output resource. The MIS or the production controller will then inform the next process about it at the given time without the original process noticing anything.

In Figure 3.10, the tiny red arrows symbolize this data exchange between devices that execute these processes with the management software. The black arrows show only the logical communication path, not the real one.

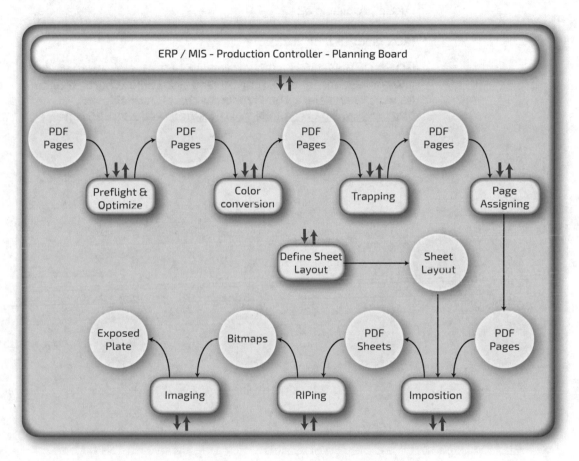

Figure 3.10: Process Resource Model for prepress in commercial printing

Reference

Hoffmann-Walbeck T., Adams R. (2020b): Visualization of the Process-Resource Model for Workflows, GRID 2020 Proceeding, p. 623-626. Available at: https://doi.org/10.24867/GRID-2020-p70 (Accessed: 15 June 2021).

Hoffmann-Walbeck T., Adams R. (2020a): JavaScript "Workflow Arrows". Available at: https://www.ryerson.ca/~wdp/workflow-puzzle/print/Test/Workflow_Arrows.html (Accessed: 15 June 2021).

Hoffmann-Walbeck T., Adams R. (2021a): Online Workflow Puzzle, Journal of Graphic Engineering and Design, Vol. 12, No. 1, p. 5-9. Available at: https://www.grid.uns.ac.rs/jged/?pid=1022 (Accessed: 15 June 2021).

Hoffmann-Walbeck T., Adams R. (2021b): JavaScript "Workflow Puzzle", Available at: https://www.ryerson.ca/~wdp/workflow-puzzle/print/print.html (Accessed: 15 June 2021).

ISO 5807: 1985, Information processing — Documentation symbols and conventions for data, program and system flowcharts, program network charts and system resources charts. Available at: https://www.iso.org/standard/11955.html (Accessed: 15 June 2021).

4 Metadata Formats

Abstract *Chapter 4 is about metadata formats in the printing industry. First, XMP (Extensible Metadata Platform) is presented. Subsequently, the Job Definition Format (JDF), the Exchange JDF (XJDF) and PrintTalk are explained in detail. The last section of this chapter is about PDF metadata relevant for print production. The basic concepts of these metadata formats are explained, not so much the actual coding.*

Keywords *JDF, XJDF, XMP, PrintTalk, PDF Metadata, Job Ticket.*

Metadata is structured data that describes the characteristics of other data (such as files). Properties of individual objects are also called metadata. Such objects could be resources or processes and much more. In this sense, the job tickets we talked about in Section 2.1.1 are examples of metadata. Other examples may describe a photo, the copyright status of a text, ISBN numbers, part numbers, serial numbers, drawing numbers, codes, and so forth. Here are four typical examples of metadata in the printing industry:

> **Definition**
>
> Strictly speaking, metadata describes properties of other data. For example, if the data represents a document that includes text and formatting rules, the corresponding metadata might contain properties such as the author's name, a copyright notice, and the creation date..

- Object descriptions ("The photo has a resolution of 350 ppi")
- Product descriptions ("Adhesive binding is required")
- Instructions ("Fold signature 1 according to F-16-6 in the fold catalog")
- Operational data ("A warning occurred during the color conversion of some content file")

In the graphic arts industry, metadata is often seen in contrast to content data (also called print data or assets). The latter describes the artworks that can be looked at on the printing substrate after printing. They are typically stored as image data, graphic data, text data, layout data, or coded in a page description language.

Metadata is either embedded in content data or stored separately from the content data, i.e. in a separate file or in a database. Most photos do not only contain the color values of pixels, but also a lot of metadata that the camera automatically generates (such as information about the camera; camera settings while the picture was taken; date, time, and GPS location). This kind of metadata is embedded, but a database might extract and store it internally in a later stage. Most content data formats allow storing metadata internally. We will cover XMP as an example of such metadata in Section 4.1 because it is occasionally used for workflow automation. If the metadata is stored externally, the data must include a reference to the content. In Sections 4.2 to 4.5, we will discuss metadata that is stored separately from the content data. In Section 4.6, I will present some less known metadata inside PDF files.

© The Author(s), under exclusive license to Springer Nature Switzerland AG 2022
T. Hoffmann-Walbeck, *Workflow Automation*, SpringerBriefs in
Applied Sciences and Technology, https://doi.org/10.1007/978-3-030-84782-1_4

Figure 4.1 shows a snippet of a sample JDF file that refers to a PDF file.

Figure 4.1:
Sample JDF data which
refers to PDF data

```
...
<RunList Run="1" Status="Draft">
   <LayoutElement>
     <FileSpec MimeType="application/pdf"
        URL="file://jobs/...sample.pdf" />
   </LayoutElement>
</RunList>
...
```

The circle in Figure 4.2 represents the metadata "hub". This is not a physical hub like those in network technology, but rather a metaphorical one. The sketch shows merely that different instances use various metadata. The print provider's production devices use metadata, but so do the print provider's business software and his partners. In a production environment for printed products, different metadata formats may be used. The data is frequently converted from one format to some other.

The most important metadata formats, of course, are JDF, XJDF, and PrintTalk. We will cover these formats in this chapter. However, as mentioned before, there are others as well. The good thing is that regular operators in a print shop don't need to worry about them. Graphical user interfaces of a workflow management system, a controller or the like obscure these technical data formats. You need to learn about these formats only if you want to extend an existing workflow, integrate a new device into the workflow

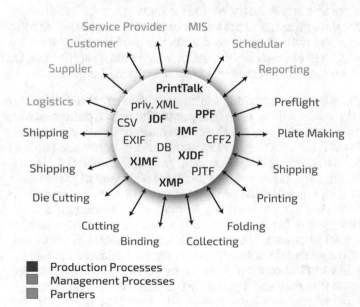

Figure 4.2:
A print provider's
metadata "hub"

■ Production Processes
■ Management Processes
■ Partners

automation network, or want to find out what is happening under the hood.

Nowadays, metadata is mostly encoded in XML or JSON. However, PPF is coded in PostScript, PJTF in PDF, CFF2 is a text format, and EXIF has its own specific binary structure. All examples in this chapter are encoded in XML; other encodings are ignored.

Figure 4.2 shows a selection of common data formats. Color profiles or preflight profiles, for instance, are not mentioned. Moreover, many companies specify private formats on their own. Why are there so many different formats? One reason is the evolution of these formats. PPF or PJTF are a bit outdated already, but they are still around, especially PPF, which is better known as the CIP3 format. Prepress often encodes ink zone presetting values in PPF/CIP3 before forwarding to press.

Another reason for the large number of metadata formats is the fact that they inhabit different ecosystems (see Figure 4.3). Some of them are prepress formats only, such as EXIF and XMP. JDF and XJDF have the widest habitat. The chart gives the impression that these formats are not being deployed between print buyer and print provider, but that is not quite true. A PrintTalk element describes business objects, which relate to products, of course. The product description inside the PrintTalk element can be either in JDF or XJDF.

The most important feature of a metadata format is what exactly it can describe. CFF2 is used for structural design in packaging, EXIF data is dedicated to image and audio data only, and XMP mainly

> **Note**
>
> **PJTF** stands for **Portable Job Ticket Format**. Adobe published the specification in 1999.
>
> **CFF2** stands for **Common File Format Version 2**. It is a text format for CAD data.

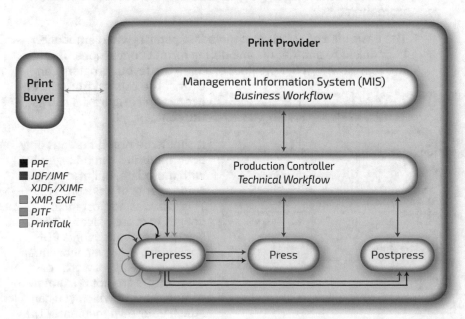

Figure 4.3: Metadata formats and their ecosystems

describes (elements of) a layout. Only JDF, XJDF and their associated messaging formats JMF and XJMF cover (almost) the entire print production scope.

In the following, we will look at the most prominent standard metadata formats only.

4.1 Extensible Metadata Platform (XMP)

Adobe Systems announced a new metadata format called *Extensible Metadata Platform (XMP)* in 2001 and released a specification in 2004. In 2008 and 2010, they split the specification into three parts. The currently valid specifications for Part 1, Part 2, and Part 3 can be found in (Adobe 2012), (Adobe 2016) and (Adobe 2020), respectively. In 2012, XMP Part 1 became also an ISO standard. The updated standard is (ISO 2019a).

Content elements such as images or graphics as well as layout files and PDF files often contain metadata entries such as copyright notices, the author's name and contact details, and keywords. They are often stored in the XMP data structure.

One application of an XMP workflow involves retrieving a caption that is registered within an image or a graphic. Often, an author delivers text and images/graphics to an agency or publisher, which then performs the layout. Traditionally, the author supplies the captions in a separate text file, and the layout artist copies them to the suitable figure. This is inefficient and, above all, error-prone. It is much more efficient if the author enters the captions in the XMP structure of the images/graphics, and the layout artist retrieves it from there.

The basic idea of XMP is that metadata persists when embedded in or linked to another file and during format conversions. This is the reason why in our previous example the layout program can extract the caption from the image/graphic. Figure 4.4 shows the concept.

It should be noted that not only document files can be tagged with metadata, but also certain components of a document. This includes, for instance, images that are placed in the layout or embedded in a PDF document. However, individual pages, paragraphs, words or even letters are not document components to which you can attach your own metadata. The

Figure 4.4: XMP metadata of a file often persist when embedded or converted into or another file.

XMP structures of the components are still "attached" to the individual components. That is, they are not merged into one common data structure. In addition, the document itself may have XMP metadata, as indicated in Figure 4.4.

As already mentioned, format conversions should preserve XMP metadata. An important application for this is the copyright entry in images. For example, when reprinting a brochure, an advertising agency may have only the PDF data that was used for printing, but no longer the image data embedded in it. Nevertheless, the agency can still extract from the PDF the XMP metadata and thus the copyright status of the images.

Unfortunately, you cannot rely on the XMP data to be maintained during a format conversion. There are two reasons for this:

- Not all data formats can store XMP. The metadata can be embedded, for example, in PDF, PostScript, TIFF, JEPG (2000), GIF, PNG and SVG files. A somewhat more complete list can be found in (Adobe 2020). The XMP data structure is an addition to a file. This means that the extensibility of a data format is a pre-condition for embedding XMP. Since the BMP image format does not have this property, for instance, it is not capable of having XMP metadata embedded.

- An XMP structure is an uncompressed XML text block within a mostly binary-encoded and compressed file (see Figure 4.6). This allows XMP reading software to easily extract the XMP data without knowing the internal structure of the file. Of course, when including an XMP structure in a binary file, the structure of the file format must still be respected. The text cannot be arbitrarily incorporated into the binary file. Figure 4.5 shows the basic integration of an XMP record into PDF. In this format, everything is strictly enclosed by PDF objects. The first object with ID 13 contains the image data. Moreover, there is a reference to some metadata object 14. Object 14 holds the corresponding XMP metadata

Accordingly, applications must support the XMP metadata format, especially when writing the structure into some data. Since Adobe originally specified XMP, it is not surprising that Adobe's applications allow XMP entries. However, many applications do not support XMP even if they generate data formats that can embed XMP.

When in doubt, test in advance if a workflow is to be based on XMP.

Note

The BMP raster data format introduced by Microsoft in 1990 does not support XMP metadata up to and including version 3.

Metadata in PDF

```
13 0 obj
<<
/Subtype/Image
Metadata 14 0 R
...
>>
stream
...(Image Data)
endstream
endobj
```

Object 12: Image Data

```
14 0 obj
<<
/Subtype/XML/
/Type/Metadata
...(XMP-Metadata)
>>
stream
...
endstream
endobj
```

Object 14: XMP Data

Figure 4.5:
Two internal PDF objects.
Object 12 is an image;
object 14 contains the
corresponding XMP
structure.

Where do the XML metadata entries actually come from? There are various options:

- The most obvious way is to enter values (for example, author's name, copyright, keywords, etc.) into a specific file manually via in an XMP-supporting application.
- Applying XMP entries in batches, for example, in all files in a specific folder. A few applications allow this approach.
- The XMP entries come from another metadata format. For example, digital cameras produce metadata about the camera and camera settings while taking the picture. The values are stored in the *EXchangeable Image File* (EXIF) format (CIPA 2016), a well-known and widely used metadata format for digital cameras. These entries are converted to XMP automatically when the photo is imported into certain image processing software, provided the software supports it.
- Photos are often stored in a database, such as in a *Digital Asset Management* (DAM) or a *Media Asset Management* (MAM) system. The database reads XMP or EXIF data from the pictures and stores the data internally in order to characterize the images. Conversely, data that is added in the database (e.g., generated automatically by means of AI image analysis) can be stored in the XMP structure of the images. Note that XMP does not have to be embedded in the files but can be saved as a separate text file. These files can also be used as a backup.
- XMP is extensible, after all the "X" in the name stands for "eXtensible". This means that companies can easily extend the XMP structure with their own metadata. For instance, image processing software can store correction steps in XMP. Most importantly, a prepress workflow management system can store internal production details.

As mentioned earlier, what makes the XMP special is its XML structure, which can be embedded in different data formats. As known in XML, the *xmlns* attribute defines XML *namespaces* and their *prefixes*. This allows new XML elements to be defined without having to worry about already existing element names. The prefixes allow identical element names as long as the prefixes are different. In Figure 4.6, the prefixes are marked in red. In the example, only reserved prefix names occur (tiff, exif, photoshop, Iptc4xmpcore). There are a few more of them. For these, the element names are fixed.

A software vendor can define its own namespace and thus its own XML substructure. Other manufacturers will usually ignore these

```
<rdf:Description rdf:about=""
   xmlns:tiff="http://ns.adobe.com/tiff/1.0/"
   xmlns:exif="http://ns.adobe.com/exif/1.0/"
   xmlns:photoshop="http://ns.adobe.com/photoshop/1.0/"
   xmlns:Iptc4xmpCore="http://iptc.org/std/Iptc4xmpCore/1.0/xmlns/"
   ...
   tiff:XResolution="3140000/10000"
   tiff:YResolution="3140000/10000"
   ...
   exif:PixelXDimension="3072"
   exif:PixelYDimension="2304"
   exif:DateTimeOriginal="2007-07-28T13:39:44+02:00"
   ...
   photoshop:CaptionWriter="Thomas Hoffmann-Walbeck"
   photoshop:Headline="Brooklin"
   ...
   Iptc4xmpCore:CiTelWork="+711-7801714"
   Iptc4xmpCore:CiEmailWork="hofmann@hdm-stuttgart.de"
</rdf:Description>
```

Figure 4.6
XMP is coded in XML

entries, but the vendor's own software can of course access the data at any time. This is a proven means of data communication within a workflow management system. As an example of such entries, I would like to mention an EAN's code, size, color, and position. For a software developer, it is easy to read XMP data from a file or to write XMP data into a file.

The XMP structure is enclosed by an XML element called *rdf*. The term stands for *Resource Description Framework*. This XML schema was adopted by the W3C in 1999 as a recommendation for storing resource metadata.

Let us wrap up: In most cases, XMP data is created to store information about a file and/or its components across applications. This supports search engines as well as contact information and copyright notices. That is, XMP is a descriptive metadata format for prepress. In the context of this booklet, however, the temporary storage of private process data is even more important. It is not standardized and therefore incomprehensible for software from different manufacturers. However, in the end, one could convert the most relevant of the private XMP data into a standard format such as JDF (after several internal process phases have been executed using XMP).

For testing purposes, you can use Adobe software to create an XMP extension without programming.

> **Definition**
>
> An **XML schema** is a description of the structure an XML document type. In particular, a schema defines the names of elements and attributes, the hierarchy of elements and the data types of attribute values.

4.2 Print Production Format (PPF)

We now come to the standard job ticket metadata formats, which are specifically specified for the description of print products and for controlling print production.

The Print Production Format (PPF) is considered the predecessor format of JDF. It was developed by the CIP3 organization, the predecessor organization of CIP4. Therefore, the PPF is often referred to as the *CIP3* format. It is encoded in the *PostScript* computer language. New structured elements have been defined to describe PPF content. They all start with "CIP3" as shown in Figure 4.7.

```
CIP3BeginPreviewImage
%%Page: 1
%%PlateColor: Cyan
CIP3BeginSeparation
/CIP3PreviewImageWidth 1490 def
/CIP3PreviewImageHeight 1210 def
/CIP3PreviewImageBitsPerComp 8 def
/CIP3PreviewImageComponents 1 def
/CIP3PreviewImageMatrix [1490 0 0 -1210 0 1210] def
/CIP3PreviewImageResolution [ 50.800 50.800 ] def
/CIP3PreviewImageEncoding /Binary def
/CIP3PreviewImageCompression /RunLengthDecode def
/CIP3PreviewImageDataSize 515348 def
CIP3BeginPreviewImage …pixels of image
CIP3EndPreviewImage
CIP3EndSeparation
       …analog for all separations…
CIP3EndPreviewImage
```

Figure 4.7: PPF snippet defining a preview image in PPF (PostScript)

The reasons why it was decided to specify JDF and JMF formats and not to try to supplement the print production format are many and manifold. In particular:

- A major shortcoming of the Print Production Format is its lack of support for order processing (MIS/ERP). PPF is used almost exclusively for technical production control.
- The PPF does not support shop floor data collection.
- There is no message format for dynamic interaction (such as the Job Messaging Format).
- PPF is coded in PostScript and therefore relatively difficult to interpret and edit.
- There is no defined mechanism for separating and merging parts of a PPF. This makes central PPF storage for different workflow components cumbersome. Instead, several differ-

ent PPF files are exchanged from and to different workflow components via various hot folders.

In essence, JDF has replaced PPF. There is only one application in which the PPF is still very popular, namely in the transfer of prepress data for the ink zone presetting in offset printing. Prepress provides the low-res image preview and transfer curves. From this, a press application can calculate the color zone preset by first applying the transfer curve to the preview, then counting the percentage of color pixels in each ink zone and for each separation. From the percentage of ink zones, the positions of the servomotors for the ink feed of the press must then be determined, depending on the paper.

Figure 4.7 shows a PPF file that describes a low-res preview image, which is suitable for calculating the ink zone presetting values. The example describes the following:

- A preview image with separations (*CIP3PreviewImageComponents*).
- The data is run-length compressed (*CIP3PreviewImageCompression*) and binary-encoded (*CIP3PreviewImageEncoding*).
- The width (*CIP3PreviewImageWidth*) and the height (*CIP-3PreviewImageHeight*) are given in pixels.
- The resolution (*CIP3PreviewImageResolution*) is given in pixels per inch (ppi).

The resolution is 50.8 ppi. The *CIP3PreviewImageMatrix* specifies the pixel direction in the image; here it is defined from left to right and next from top to bottom. The pixel values are stored inside the PPF after the *CIP3PreviewImage* element.

4.3 Job Definition Format (JDF)

The introduction of JDF and the associated *Job Messaging Format* (JMF) by the *Cooperation for the Integration of Processes in Prepress, Press and Postpress* (CIP4) in the year 2000 was a major innovation. Through the subsequent JDF implementations, workflow automation in the print production has increased significantly.

First, let's take a look at the simple example of JDF integration in Figure 4.8. The MIS initially records quite a lot of data about the print product and its production during the estimating of a quote. If this estimation converts to a print job, some of the data will eventually be sent to prepress, press, and postpress, such as details about the intended product and some business data (such as customer details, delivery date, etc.). First, prepress will process

> **Definition**
>
> A **transfer curve** is a curve/table that defines modifications of tonal values of the artwork during RIPing to match desired tonal value increases in print – for example, to reach an offset standard.

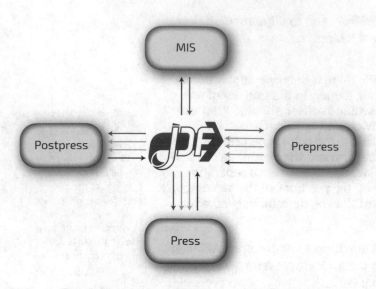

■ Order ID, Customer Details, Page Size, Sheet Size, Colors, Printing Substrate, Run Length, Imposition Scheme,...
■ Ink Zone Prestting Values
■ Previews
■ Cutting Data, Folding Data
■ Operational Data

Figure 4.8: Example of information exchange with JDF

the JDF and add new entries, such as ink zone presetting values for the press. The RIP might also generate composite previews for press and postpress, for example, for quality assurance purposes. Moreover, the imposition software might forward cutting and folding positions to the guillotine and the folder. Finally, all these devices should send operational data back to the MIS.

We'll talk about the distribution of JDF later. For now, it should be sufficient to have a JDF pool of some sort as a hub.

What do the arrows in the Figure 4.8 are present? Note that the information content can vary regarding its details. For instance, if only the order number is transmitted, the level of detail is low. If, however, many production details are transferred so that a subsequent process can run fully automatically, the level of detail is high. All JDF interfaces usually operate between these two extreme positions.

This circumstance also explains why workflow managers in print shops and at manufacturers should be well versed in interface details. The benefit of JDF integration depends strongly on the level of detail in the JDF data.

Before discussing the details of JDF and JMF data structures, it should be mentioned that JDF workflow systems are usually configured very individually for a print shop. Since JDF workflow systems tend to affect many areas of a print shop, existing components are usually integrated.

However, JDF and JMF are merely standardized data formats and not specifications for workflow systems. This means that components from different manufacturers communicate with each other via JDF/JMF but otherwise actually act independently of each other. This can lead to incompatibilities that must be solved case by case (see Section 4.6). In particular, this means that JDF/JMF integration is not an off-the-shelf software package that can be purchased and simply installed, but a project that must be pursued in several stages and over the long term.

Consequently, setting up a JDF workflow system requires a great deal of commitment, expertise and project management skills.

In a JDF-based workflow, a specific JDF file is created for each print job, sometimes in multiple versions as production progresses. As Figure 4.8 suggests, the MIS often writes the first JDF file. Usually, it contains a description of the intended product and rough production definitions without much detail. As processes add additional data to the JDF during production, the file size will consequently increase.

4.3.1 JDF Nodes

JDF and JMF are XML-encoded. That is, each JDF or JMF record consists of a tree of XML elements and their attributes and values. The permitted XML element names are defined in the JDF specification (CIP4 2020). Since JSON has become increasingly popular in the last decade, CIP4 is currently working on a JSON representation of JDF/XJDF.

First, let us talk about the XML elements for JDF nodes. They define the product to be manufactured and, if applicable, its sub-products, as well as the production processes needed to manufacture this product.

Each JDF node has the attribute *Type*. The value of this attribute can be either *Product*, *ProcessGroup*, *CombinedProcess*, or some specific process name such as *ImageSetting*, *ConventionalPrinting*, or *Folding*. CIP4 defines over 100 different processes.

For a standard JDF file, the root JDF node has the value *Product*

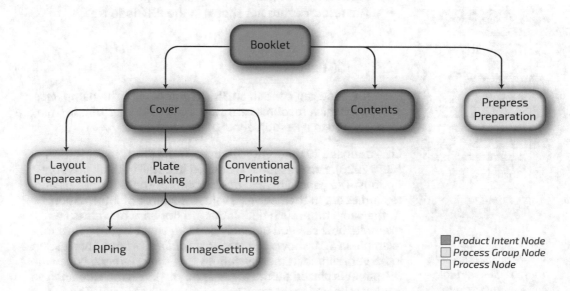

Figure 4.9 Example of a JDF Node tree for the production of a booklet

for the attribute *Type*. This node is called a *Product Intent Node*. It represents the final product. Underneath the root, there can be more *Product Intent Nodes*. These represent sub-products such as cover, content, and the like. Below the *Product Intent Nodes* there are *Process Group Nodes* and *Process Nodes*. They describe for each sub-product the processes that are required for its production. Please note:

- The node hierarchy is not limited. Figure 4.9 shows merely an example.
- As usual, the hierarchy of an XML tree is defined by creating subnodes. That is, the root node in our example contains two JDF child nodes of type *Product*, which in turn contain further child nodes. A *Process Node* is always a leaf of the tree (i.e., it does not contain any further subnodes), while *Process Group Nodes* can optionally contain further subnodes (further *Process Group Nodes* or *Process Nodes*).
- The *Process Nodes* and the *Process Group Nodes* below some *Product Intent Node*, which represents a product part, refer only to the production of this product part.
- Not all processes that are needed to produce a print product need to be listed in the JDF file. In general, only those processes which are included in the JDF workflow will be defined.
- The JDF file reflects only the current situation at a given point in production. In general, more nodes may be added as production progresses.
- In a real production environment, the node tree is often much more complex than in this example.
- The resources are not shown in the JDF node tree.

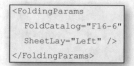

```
<FoldingParams
  FoldCatalog="F16-6"
  SheetLay="Left" />
</FoldingParams>
```

Figure 4.10
Resource FoldingParams

4.3.2 JDF Resources

Each JDF node can contain an XML element with the name *ResourcePool*, in which resources are stored. All resources of a JDF file are incorporated in a ResourcePool of some node.

CIP4 defines 170 different resources in the latest specification (CIP4 2020). Each resource is defined in detail with all possible options. The generally valid descriptions of these processes and resources are in themselves a valuable source of information. At the same time, such definitions are not easy to create. For example, how can you uniquely define a sheet layout? The code example in 4.10 shows the resource *FoldingParams*. This resource looks very simple. It merely defines that the reference edge where the paper is placed on the folder is on the left-hand side and the folding scheme is according to catalog number F16-6. The real

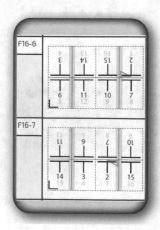

Figure 4.11:
Two folding schemes from the CIP4 fold catalog. The catalog contains almost 100 schemes.

achievement was to define a unique fold catalog. This catalog can be found in the JDF specification (CIP4 2010). It includes all common folding schemes, almost 100 different ones in total. Figure 4.11 shows the folding schemes F16-6 and F16-7.

The specification of the folding catalog number does not suffice for an automatic folding machine preset. In the *FoldingParams* resource, the exact folding position can optionally be specified. Only then can a folding machine preset itself accurately according to the specifications in prepress (ignoring the paper distortion during printing). Again, this shows how important it is to know the information details of interfaces.

Input resources of a Process Node or a Process Group Node are the physical items such as plates, electronic items such as files, or conceptional items such as a parameter set that a process or process group needs to observe. An output resource is an item that the process or process group generates during execution. That is, a process consumes input resources and produces output resources. Thus, JDF is based on the process-resource model that we dealt with in Chapter 3. But what about *Product Intent Nodes*? Do they have input and output resources as well?

The answer to this is yes. The interpretations of those resources differ, though. The output resource of a *Product Intent Node* is always called *Component* and represents the final product or a product part - similar to the node itself. A Component can actually be an input resource for another Product Intent Node one level up in the hierarchy. For example, the Component that represents a cover is an input resource of the Product Intent Node representing the final booklet. The input resources of a Product Intent Node, which is not transitional (see Section 3.3, i.e., is not an output node of another Product Intent Node) is called *Intent Resource*. An Intent

> **Definition**
>
> A JDF **Component** describes the various versions of (semi-) finished goods. The final product, product parts, but also a set of printed sheets or a pile of folded sheets are considered components.

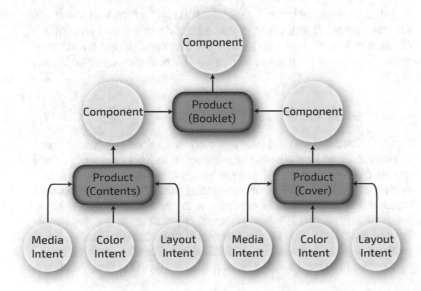

Figure 4.12: Example of Product Intent Nodes and their input and output resources

Resource specifies the print buyer's intention for the product (part), i.e., details of the product (parts) without defining the processes needed to make them. Figure 4.12 explains the terminology. Note that here the arrows do not represent relationships between JDF nodes as before, but the relationships between resources and nodes. The *MediaIntent* defines the printing substrate, the *ColorIntent* the separations for this job part, and *LayoutIntent* the number and dimensions of the pages.

In the remainder of this section, we will talk about *Partitionable Resources*. In general, one can talk about things in their entirety, as certain subsets, or as singularities. For example, one can talk about books in general, about certain categories of books, or about a particular book. This is similar to some resources for a print job. For example, the resource *ExposedMedia* specifies exposed printing plates (among other things). You may want to specify all plates for a print job or only a subset, i.e.:

- All plates of the job
- All plates of a signature (of the job)
- All plates of a sheet (of a signature and the job)
- all plates of a side (of a sheet, a signature, and the job)
- A specific separation (of a side, a sheet, a signature, and the job)

These kinds of resources are called *Partitionable Resources* in JDF terminology. The specification details how certain resources can be partitioned. Note that the partitioning may not always be identical for different resources. For instance, print sheets cannot be partitioned up to the separation, as was previously the case with the printing plates.

Another example is the printing substrate, which can be the same for the entire job, but often differs based on cover and content. Of course, the content can also be printed on differing substrates. With the help of partitions, all this can be defined easily and flexibly.

4.3.3 Structure of a JDF File

Product descriptions and processes are coded via JDF nodes. We already learned that resources are pooled in XML elements called *ResourcePools,* and that those elements are sub-elements of JDF nodes. Thus, every JDF node, no matter what type, can optionally contain a *ResourcePool*. However, not every resource that is either input or output of a JDF node necessarily resides in the *ResourcePool* of that node. The reason for this is that resources can have relationships with multiple JDF nodes, in particular transitional resources. In order to avoid having to keep (and update) resources in multiple copies in different JDF nodes, JDF nodes may also have input or output resources which lie outside of the particular JDF node.

The logical consequence of this is that there must be some sepa-

```
<?xml.....?>
<JDF Type="..." ...> ...
        <ResourcePool>...
        </ResourcePool>
        ...
        <ResourceLinkPool>...
        </ResourceLinkPool>
        ...
        <JDF Type="..." ...>
                ...    other JDF sub node(s)
        </JDF> ...
</JDF>
```

Figure 4.13:
Structure of a JDF file

rate information that specifies which resources a JDF node is linked to. This information is called *ResourceLink*, where all *ResourceLinks* of a JDF node are collected in a *ResourceLinkPool*. A *ResourceLink-Pool* is a subelement of a JDF node. This results in the JDF file structure described in Figure 4.13.

4.3.4   Audit Pool

In Figure 4.13, an important pool in the JDF node is missing. In addition to the *ResourcePool* and the *ResourceLinkPool*, a process node can also contain an *AuditPool*. *Audit Elements* are written to this pool so that the process results can be recorded after execution. Typical contents of an *Audit Element* are:

- Generation, modification or deletion of a JDF node
- Process times (Start, End, etc.)
- End status (Completed, Aborted, Stopped, etc.)

```
<AuditPool>
    ...
    <ProcessRun
        AgentName=    "CIP4 JDF Writer Java"
        End=          "2020-12-05T13:53:52+01:00"
        EndStatus=    "Completed"
        Start=        "2020-12-05T13:53:43+01:00"/>
    ...
</AuditPool>
```

Figure 4.14:
Code snippet of a ProcessRun Audit Element in an AuditPool

- Errors (Warning, Fatal, Error, etc.)
- Reaching of a milestone
- Consumed or missing resources

With the contents of the *AuditPool*, an MIS can post-calculate the print job, for example. Figure 4.14 shows a small code excerpt of an *AuditPool*.

The attribute *AgentName* holds the name of the application that added the Audit Element to the *AuditPool*. Beside the shown *Audit Element* of type *ProcessRun*, there are others, such as:

- *PhaseTime* for logging start and end times of any process states and process phases
- *Notification* for logging events (such as errors) during process execution
- *ResourceAudit* for logging the usage of resources during execution
- *Created, Modified,* and *Deleted* for logging the creation, modification, and deletion of a JDF node or a resource

4.3.5   Gray Boxes

As mentioned before, an MIS normally cannot describe details of the production, only a rough framework. To describe such framework, special Process Group Nodes have been specified. They are called *Gray Boxes*. A Gray Box does not define all processes or all resources for the embraced processes, except for the (final) output resources. The information contained in a Gray Box does not suffice to execute the processes that the Gray Box holds, which is why a Gray Box is called non-executable. The missing data must be added as the production progresses. If all necessary data is available, the Gray Box is dissolved and may become a normal Process Group, for example.

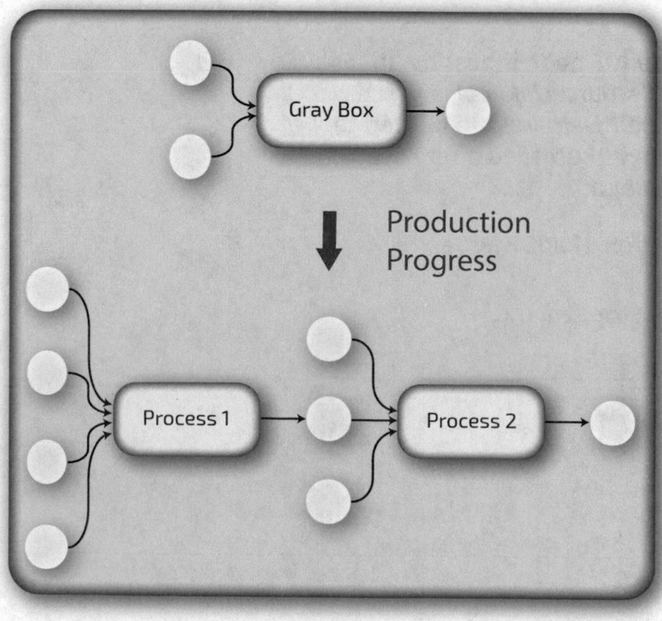

Figure 4.15: A Gray Box must become (a group of) processes with all mandatory entries and resources before it can be executed.

70

You can recognize a Gray Box by the fact that it

a) is a process group, i.e. *Type="ProcessGroup"*, and
b) has the attribute *Types* (note the "s"!).

The value in the *Types* attribute is normally an (incomplete) list of processes or sometimes a predefined name. The process list gives an indication of which processes (among others) the Gray Box should be resolved to. The predefined names are specified in the *ICS* papers (see Section 4.6). The Gray Box name *PrePressPreparation* would be such an example. Further options and details are specified in the *ICS* papers.

Note

ICS stands for Interoperability Conformance Specification.

4.3.6   Spawning and Merging

In a print shop, products are produced by parallel or overlapping processes. For example, layout elements such as text, images and graphics are created in parallel (in JDF terms: *LayoutElementProduction*). An example of overlapping execution is that while the set of plates for a print product is still being imaged, some of the previously exposed plates are already used for printing.

This implies that the JDF structures (nodes and resources) of a JDF file must be sent to different controllers/devices simultaneously. This would not be a problem in itself, but keep in mind that the devices (should) provide the *AuditPools* with fresh audits. Resources might also be updated. For instance, only after RIPing can values for the ink zone presetting be calculated and stored in the corresponding resource. However, this can cause conflicts when the modified JDF structures are merged back into the original JDF file. This is JDF technology has a process called *"Spawning* and *Merging"*. This mechanism requires that it must be recorded in the original JDF file if some part of the JDF structure is copied for a controller/device (*Spawning*). Whether the recipient has read-only or also write permissions is also important. A JDF structure may be checked out with write permissions only once at any given time. If the modified structure is merged back into the original file, it should be recorded once again. The JDF structure may then be spawned once more with full write permissions. This way, there are no more conflicts when merging. Figure 4.16 shows the principle of this mechanism.

However, if two production lines operate in parallel, the processes are more likely to be modeled completely separately. For example, the cover and the contents of a brochure can be printed simultaneously on two different offset presses. This situation will be modeled via two different *ConventionalPrinting* processes.

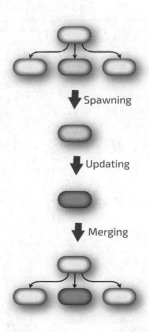

Figure 4.16:
Spawning and merging of a JDF node

4.3.7   Job Messaging Format (JMF)

Job Definition Format (JDF)

❶ *(Parts of a) Job Ticket*

❷ *Job Ticket Updates, Audits*

Job Messaging Format (JMF)

❸ *Commands, Queries*

❹ *Response, Signals*

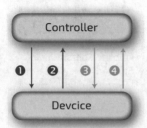

Figure 4.17:
Key data exchange
between controller and
device

Now that we have discussed what kind of information can be stored in a JDF file, the question naturally arises: How do workflow instances exchange this data? In the simplest case, JDF job tickets are written by a JDF producer (see Figure 4.17-❶) and placed into the hot folder of a JDF consumer, who scans the folder for new data in certain intervals. This unidirectional interface via file transfer is, of course, quite static, slow, and does not allow real-time interaction. It also can become complex to maintain if many different workflow instances are involved. Real-time interaction, however, is necessary, if you think about Job tracking, job changes, instant error messages, and the display of actual material consumption or the degree of machine utilization.

Moreover, one should give up on the idea that a JDF file is copied to the input queue (or hot folder) of a controller/device via file transfer. The system sends a message containing the location of the JDF file instead (see Figure 4.17-❸). But how should one imagine such a message?

The Job Messaging Format (JMF) is part of the JDF specification. Like JDF, JMF is also XML-encoded. In the case of JMF, the root element is no longer JDF but rather JMF. JMF messages use the HTTP and HTTPS protocols, i.e., the JMF is the body data of an HTTP(S)

```
<JMF TimeStamp="..." SenderID="ID4711">
  <Command ID="M1" Type="SubmitQueueEntry">
    <QueueSubmissionParams>
      URL="File://Computer/Directory/job.jdf"/>
  </Command>
</JMF>
```

Figure 4.18: JMF command informing the recipient where to retrieve a JDF file

packet. The JMF code in Figure 4.18 shows an example of a JMF message. It contains a *Command* telling the recipient the location of the submitted JDF file *job.jdf*. Analogously, the recipient can also be told how to reach the JDF data via the HTTP protocol instead of via file transfer. More generally, numerous control messages concerning queues and their entries are important JMF applications. Others are:

* System bootstrapping and setup – for example, shutting down or waking up a device (which is in standby mode).
* Dynamic status, resource usage and error tracking for jobs and devices.
* Pipe control for overlapping processes. One process produces

resources while at the same time some other process consumes portions of these resources that are already available (see Section 2.4.5).

- Device setup and job changes – for example, creating new JDF nodes, which might be necessary if a workflow controller receives the artwork for a job but the MIS has not yet defined a JDF job ticket. The workflow controller will then send a *NewJDF* command to the MIS to initiate a new job. (See Section 2.2.1)

- *Device Capability* description for sending technical capabilities of a device to a controller, such as the minimum and maximum sheet size of a sheet-fed press.

Describing device capabilities can be quite tricky, though. The device capability might not be a fixed set of properties. Some machines (for example, a folding machine) can be optionally equipped with additional units, which changes the properties. In addition, properties can vary due to other factors. For example, the possible number of folds that a folding machine can perform depends on the paper thickness.

From a more abstract point of view, six different kinds of messages are available (called *JMF Message Families*):

- *Query*
- *Command*
- *Response*
- *Acknowledge*
- *Signal*
- *Registration*

A *Query* sends an inquiry to a recipient without changing its state, for instance, to retrieve information about a device's status or material consumption. A *Command*, however, changes the state of the addressee. An example is the deletion of a job in a queue or the submission of a JDF job ticket (see Figure 4.18). Both *Queries* and *Commands* require the receiver to send a *Response* message back to the sender. If a command (or query) takes a while to execute, it may be necessary to send a separate and asynchronous *Acknowledgment* message back to the sender after completion. A *Signal* is a unidirectional message, mostly from a device to a controller, usually about a status change. Signals are often forwarded automatically as "fire and forget" messages; that is, no responses are required. With a *Registration,* a controller can subscribe to certain signals from a device or another controller automatically, for instance, in certain time intervals. A controller may also request a subscription on behalf of others. For example, an MIS might send a *Registration* to a prepress WMS in order to make sure that signals are sent to a press controller whenever a plate has been produced.

JMF messages are not only quite diverse but can also occur in large quantities. Therefore, let us conclude this section with the statement that JMF messages are mostly used for dashboards that visualize the current state of the shop floor. An MIS, on the other hand, uses the already summarized results in the *AuditPools* for post-calculation.

4.4 Exchange Job Definition Format (XJDF)

The JDF concept is over 20 years old already. Back then, the average run-lengths of jobs were much higher than today, and the average number of print jobs processed each day in a print shop was much lower. With a few dozens of print jobs per day, storing a separate JDF file for each print job is feasible. Today, however, a large online printer may run 10,000 jobs per day. Loading all these files into a workflow system in order to display the current jobs and their actual status would take much too long. Thus, a workflow controller such as an MIS operates a private database to speed up importing job data. This approach has been state-of the art for quite a while already. At the same time, the JDF handling has become more centralized. Either the MIS or a workflow controller is in charge of the JDF job tickets. This JDF master controller has many tasks. It must monitor hot folders and HTTP ports for incoming data. It must also send out data to other controllers and devices. To do this, it must monitor spawning and merging. It would be JDF compliant if a complete JDF file were transmitted to a device or another controller. Each device must be prepared to retrieve the required data from the file on its own, if necessary. However, this would be unfavorable since the complete JDF data set would be "blocked". Therefore, it is more efficient if the software sends only those JDF parts to a device that it needs. Thus, the master controller must be able to pick out the relevant parts for each device (usually one or more JDF nodes and the corresponding resources). In this case, JDF is a communication protocol between controller and devices. The workflow logic will be stored partly in the JDF files and partly in the controller's private database. Often, the workflow logic is merely part of the MIS and systems use JDF as a pure information interchange technology.

In 2018, the new Exchange JDF (XJDF) was published. The current version is from 2020 (CIP4 2020b). This new format reflects the changes in the JDF workflow implementations that we just discussed. XJDF is not an electronic job ticket like JDF but rather an interface specification. It is merely a protocol between two workflow components, primarily between controller and devices. It does not contain the workflow logic anymore, as JDF did. The workflow

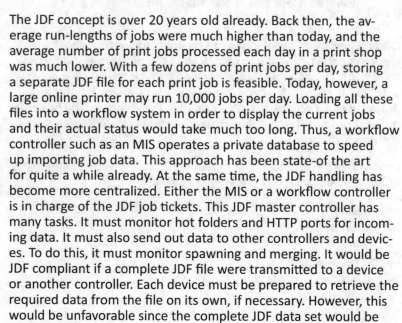

Definition

The **workflow logic** (also called **process logic**) is the description of rules and relationships between processes and resources that are needed for print production. In other words, it is the JDF's underlying process-resource model.

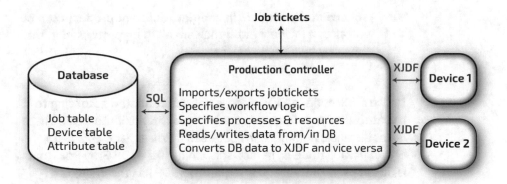

Figure 4.19: Simplified XJDF architecture

logic is now hidden in the internal data structure of the workflow system. This workflow system could be an MIS or some production control system. The internal data structure is undisclosed.

At first glance, it seems as if nothing of the "old" JDF has remained after the redesign. This is deceptive, though. The descriptions of the processes and resources have essentially remained the same. Only the connections between the processes (using the transitional resources) are no longer mentioned in XJDF. Moreover, there is no complete description for the print production at the end – at least not in XJDF. In fact, XJDF is only generated at runtime, then forwarded and read. Afterwards, it is discarded. There is no XJDF storage.

As a result, XJDF is much less complex:

- Since the XJDF addresses a single device only, an XJDF node has no children. Therefore, the nested node tree we saw in JDF no longer exists (see Section 4.3.1).
- All resources for a XJDF node are embedded in the node. There is no *ResourcePool* anymore. Similar resources are compiled in a so-called *ResourceSet*. There might be more than one *ResourceSet* in an XJDF node. Since all resources are inside the XJDF node, there is no longer the need to define *ResourceLinks*, and more searching for a resource in the node tree using *ResourceLinks* (see 4.3.2).
- Change orders can be implemented by simply updating a XJDF node and sending it once again.
- *Spawning* and *merging* has been removed (see 4.3.6). In JDF, *Process Nodes* and *Process Group Nodes* are subnodes of a *Product Intent Node*. Moreover, a JDF file is only valid for one product (and its sub-products). In XJDF, the product description is not directly connected with the production details anymore. Thus, changing the product structure (adding another product part, for instance) does not affect the

production description. In addition, different products can be written to a *Product List* if they are all to be processed in the same way by a single device.

Unlike a JDF node, a XJDF node is not differentiated according to the attribute *Type*. It does not even have this attribute at all. There is, for instance, no difference between a *Process* and a *Process Group*. A XJDF node is, in fact, something like a JDF Gray Box, that is, it can be incomplete. Consequently, XJDF has the *Types* attribute, whose values can hold either one or more process names or the word *Product* (or both). Unlike a JDF Gray Box, however, an XJDF node does not need to be expanded with additional information before it may be submitted to a device for execution. If an incomplete XJDF node is submitted, the device is expected to fill in the missing information on its own with default values. In fact, almost all XJDF attributes are optional.

```
<?xml.....?>
<XJDF Types="Process Name(s)" ...>
       ...
       <AuditPool>...</AuditPool>
       <ProductList>...</ProductList>
       <ResourceSet>...</ResourceSet>
            ...
       <ResourceSet>...</ResourceSet>
</XJDF>
```

Figure 4.20: Structure of XJDF data

Figure 4.20 shows the structure of XJDF data. As mentioned above, it contains only a single XJDF node. Since JDF and XJDF are structurally different, XJDF is not backward-compatible with JDF. However, an XJDF element can be automatically converted to a JDF element and vice versa. Of course, XJDF data cannot be converted into a JDF job ticket containing the complete workflow logic.

XJDF coexist with JDF. The parallel development of both versions by CIP4 offers options to industry stakeholders wishing to incorporate JDF, XJDF or both into their product offerings and production environments. However, several print automation experts believe that XJDF is the format of the future and will gradually replace JDF.

Figure 4.21 shows a code snippet from an XJDF node that describes the product only from the print buyer's point of view, not the production. The value of the Types attribute is therefore Product. The code could, for example, come from a web-to-print system (B2C) or from any other external site (a subsidiary, for instance)

```xml
<?xml version="1.0" encoding="UTF-8" standalone="yes"?>
<XJDF Category="Web2Print" JobID="4711" Types="Product">
  <ProductList>
    <Product Amount="300">
      <Intent Name="MediaIntent">
        <MediaIntent MediaQuality="Lumisilk" MediaType="Paper"/>
      </Intent>
      <Intent Name="LayoutIntent">
        <LayoutIntent FinishedDimensions="240.0 155.0 0.0"
         Pages="1" Sides="OneSided"/>
      </Intent>
      ... more Intents
    </Product>
  </ProductList>
  <ResourceSet Name="RunList">
    <Resource>
      <RunList>
        <FileSpec URL="asset/Cool.pdf"/>
      </RunList>
    </Resource>
  </ResourceSet>
    ... more ResourceSets
</XJDF>
```

Figure 4.21: XJDF Node of Types="Product"

and be addressed to the MIS. In the following section 4.5 we will learn that such XJDF codes can also be embedded into a PrintTalk element.

The XJDF node includes a product list, but it contains only a single product in our example. 300 copies are requested. Intent sub-elements describe the technical details of the product, similar to the Intent Resources in JDF. Only *MediaIntent* and *LayoutIntent* are listed here. *MediaIntent* defines the substrate while *LayoutIntent* defines the size (*FinishedDimension*) of the final product, the number of pages (*Pages*), and the number of printed surfaces (*Sides*). The size is given in DTP points, where the first value specifies the width and the second value the height. A *ResourceSet* describes a set of one or more Resource elements of the same kind (see also Figure 4.25). In Figure 4.21 there is only one Resource *RunList* that specifies the storage location of the content file.

The XJDF node in Figure 4.22 is a protocol that travels from a controller to a folding machine (*Types="Folding"*). This example is a flyer to be produced 4/4 in a run of 1,000. In *ResourceSet FoldingParams*, only the desired fold catalog number is specified. This catalog was already presented in 4.3.2.

If you want to learn more about XJDF, I would like to recommend

Definition

A **RunList** defines one or more printable logical documents, e.g., a PDF file.

(Meissner 2017). I would also like to mention the *EasyXJDF* tool written by the same author (Meissner 2018), which lets you generate simple XJDF examples that could be used for W2P.

```
<XJDF JobID="Job1234" Types="Folding"...>
  <AuditPool>
   ...Audits...
  </AuditPool>
  <ProductList>
    <Product Amount="1000" DescriptiveName="Flyer">
      <Intent Name="ColorIntent">
         <ColorIntent>
            <SurfaceColor Surface="Front"
              ColorsUsed="Cyan Magenta Yellow Black"/>
            <SurfaceColor Surface="Back"
              ColorsUsed="Cyan Magenta Yellow Black"/>
         </ColorIntent>
      ... More Intents
    </Product>
  </ProductList>
  <ResourceSet Name="FoldingParams" Usage="Input">
    <Resource>
      <FoldingParams FoldCatalog="F2-1"/>
    </Resource>
  </ResourceSet>
  ... More ResourceSets
</XJDF>
```

Figure 4.22: XJDF Node of Types="Folding"

4.5   PrintTalk

PrintTalk is an XML-based, open data format used to describe commercial/business activities in the graphical industry. It is used mainly as an interface between the print buyer and the print provider. As an example, let's look at the interface between an external W2P system and a print provider's MIS/ERP system. The second area of application is currently the interaction of print brokers, subcontractors and branch offices with a print provider. We discussed this in Section 2.2.

The root element in a PrintTalk document has the name *PrintTalk*. There are two important sub-elements, the *Header* and the *Request* (see Figure 4.23). The *Header* identifies the original sender and the recipient of the PrintTalk transaction. The *Request* is just a container for a *Business Object*.

There are 15 different Business Objects altogether (ordered alphabetically):

> **Note**
>
> Initially, 16 companies have announced a project named **PrintTalk**. Next , NPES (now *Association of Print Technologies* or *APTech*) organized and published the PrintTalk specification. In 2005, the PrintTalk development was transferred to CIP4 for its long-term maintenance and distribution.

Cancellation, Confirmation, ContentDelivery, ContentDeliveryRe-sponse, Invoice, OrderStatusRequest, OrderStatusResponse, Proo-fApprovalRequest, ProofApprovalResponse, PurchaseOrder, Quotation, Refusal, RFQ, StockLevelRequest, StockLevelResponse

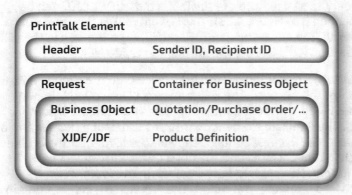

Figure 4.23:
*Structure of a PrintTalk
Element*

In Figure 2.3 we have already seen the classic business workflow, which starts on the print buyer's side with an *RFQ* (Request for Quote), followed by a *Quotation* (PP->PB), *Purchase* Order (PB->PP), *Confirmation* (PP->PB), *ContentDelivery* (PB->PP), *ContentDelivery-Response* (PP->PB), *ProofApprovalRequest* (PP->PB), *ProofApprov-alResponse* (PB->PP), and *Invoice* (PP->PB). This workflow is drawn in more detail in Figure 3.3, which also indicates possible refusals of business objects, for example, if the PP rejects an RFQ or the PB rejects a quotation.

In (PrintTalk 2020c) you can find PrintTalk workflows (Figures 6.1 to 6.8). Two of them are shown in Figure 4.24. In workflow ❶, the print buyer queries the print provider about status of his order us-ing *OrderStatusRequest.* This business object can be defined either for one single or for multiple status requests. It can also specify if there should be a unique *OrderStatusResponse* or if the request is supposed to be interpreted as a subscription. In this case, an *OrderStatusResponse* message should be sent whenever a new milestone is reached.

Workflow ❷ represents a con-venient workflow for B2B. If the PP maintains a warehouse for a PB, the PB can use the *Stock-LevelRequest* business object to query whether and how many products are still available in the warehouse. The PP should return a *StockLevelResponse* containing the number of items that are currently available. He

> **Note**
>
> PrintTalk is a protocol on top of HTTP. HTTP and protocols further down in the protocol stack (like TCP/IP) are responsible for the actual network routing of the PrintTalk element. The identifi-cation in the PrintTalk header is ensured by means of the Web URLs and the DUNS numbers of the companies in-volved. DUNS stands for Data Universal Number-ing System. It is a unique nine-digit identifier for businesses.

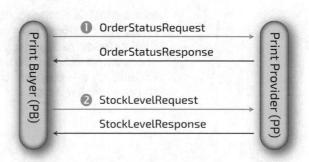

Figure 4.24:
Two different PrintTalk workflows between PB and PP

can also return the price of the item. Moreover, he can communicate if new copies are planned for production in case no goods are currently available in the warehouse.

The very special feature of PrintTalk is that XJDF or JDF can be embedded (see Figure 4.23). This is what makes this protocol so powerful. The embedding of JDF and XJDF is described in the (CIP4 2015a) and (CIP4 2020c) specifications, respectively.

Let us take a closer look at the PrintTalk element example in Figure 4.25. For this purpose, the key terms are color-coded (element names in red, attribute names in green, and attribute values in blue). The PrintTalk element consists, as usual, of a *Header* and a *Request*. The *Request* contains the *PurchaseOrder* business object, which in turn encloses an XJDF object. This XJDF object consists of a *ProductList* and two *ResourceSets*. In the *ProductList*, the purchased product is described. The *ProductList*, in fact, contains three products. The first, with the product type *Booklet*, represents the final product. It includes two *Intents*, one for the *BindingIntent*, the other for the *LayoutIntent*. The *BindingIntent* defines, in particular, the *ChildRefs* attribute. The associated values are the IDs of the following two semi-products, representing the cover and the content of the booklet. Both include the *MediaIntent* and *LayoutIntent*, which determine the paper (*MediaQuality*) and the number of pages (*Pages*), respectively.

One *ResourceSet* defines the customer's *Contact* details, the other one specifies the PDF files containing the artwork. The latter information is located in two *RunList* resources. The first PDF file named *Cover.pdf* has two pages (*NPage*), the second named *Content.pdf* has eight pages.

The example in Figure 4.25 is a somewhat simplified version of the original file. For example, the prefix *ptk* for all PrintTalk element names is present, while I omitted the prefixes for the XJDF elements to make the example a bit easier to read. Moreover, I removed a few attributes, such as *Amount* in the Product elements, which specifies the desired number of copies.

Many PrintTalk Elements do not contain a *ProductList*. The business object *ContentDelivery*, for example, only needs to enclose a *RunLIst ResourceSet*, specifying the file path and name of the delivered content file. The delivery must relate to some purchase order, though. For this purpose, the attribute *BusinessRefID* is provided in the element *Request*.

On the next page – Figure 4.25:
Example of an XJDF element

```xml
<?xml version="1.0" encoding="UTF-8" standalone="yes"?>
<ptk:PrintTalk Version="2.1" Timestamp="2019-09-09T16:27:38Z"
      xmlns:ptk=http://www.printtalk.org/schema_2_0
      xmlns="http://www.CIP4.org/JDFSchema_2_0">
  <ptk:Header>    Information about Sender and Recipient
  </ptk: Header>
  <ptk:Request BusinessID="ID4711">
    <ptk:PurchaseOrder Expires="2021-05-12T19:26:41Z">>
       <XJDF JobID="ID4712" Types="Product">
        <ProductList>
          <Product ID="ID_broschur" IsRoot="true" ProductType="Booklet">
            <Intent Name="BindingIntent">
              <BindingIntent BindingOrder="Collecting" BindingSide="Left"
              BindingType="SaddleStitch" ChildRefs ="ID_Cover ID_Content"/>
            </Intent>
            <Intent Name="LayoutIntent">
               <LayoutIntent NamedDimensions ="A4"/>
            </Intent>
          </Product>
          <Product ID="ID_Cover" IsRoot="false" ProductType="Cover">
            <Intent Name="MediaIntent">
              <MediaIntent MediaQuality="P1" MediaType="Paper"/>
            </Intent>
            <Intent Name="LayoutIntent">
              <LayoutIntent Pages="4"/>
            </Intent>
          </Product>
          <Product ID="ID_Content" IsRoot="false" ProductType="Content">
          <Intent Name="MediaIntent">
            <MediaIntent MediaQuality="P2" MediaType="Paper"/>
          </Intent>
          <Intent Name="LayoutIntent">
            <LayoutIntent Pages="8"/>
          </Intent>
          </Product>
        </ProductList>
        <ResourceSet Name="Contact" ProcessUsage="Input">
          <Resource ID="Contact_123">
            Contact ContactType="Customer">
              <Person FamilyName="Duck" FirstName="Donald"/>
            </Contact>
          </Resource>
        </ResourceSet>
        <ResourceSet Name="RunList" ProcessUsage="Input">
          <Resource>
            <Part Run="R_Cover"/>
            <RunList NPage="2" Pages="0 1">
               <FileSpec URL="File:///dir/Cover.pdf"/>/>
          </RunList>
          </Resource>
            <Resource>
             <Part Run="R_Content"/>
             <RunList NPage="8" Pages="0 7">
                <FileSpec URL="file:///dir/Content.pdf"/>/>
            </RunList>
          </Resource>
        </ResourceSet>
      </XJDF>
    </ptk:PurchaseOrder>
  </ptk:Request>
</ptk:PrintTalk>
```

4.6 Interoperability Conformance Specification (ICS)

Without a doubt, a JDF/XJDF-compatible system cannot be connected via plug-and-play. Manufacturers test interfaces in advance. An integration matrix, which can be found on the CIP4 website (cip4.org), provides an overview of products that have been tested and integrated with each other.

The reasons for incompatibilities are of different nature. First, let us look at the fact that JDF and XJDF element names, attribute names and attribute values can be privately extended. This is to allow a manufacturer to exchange private information between its system components based on JDF/XJDF. Such extensions are ignored by systems from third-party manufacturers. Of course, companies are urged to not replace standard JDF/XJDF keywords with their own, but at best to only supplement them. If this is not properly observed, it can lead to incompatibilities.

The main reason for incompatibilities is probably a different one, however. In general, languages are complex systems of communication. This is true not only for a language between humans, but also for computer languages and even data formats such as JDF. You can express an issue differently, even if you follow all the official rules, the grammar, and the official vocabulary. RIPs might render a PDF file differently. Similar is true for JDF. Some people say that different JDF dialects have emerged. This terminology, however, is a bit problematic, because a dialect is a modification of the grammar and/or vocabulary of a language. However, two JDF systems may not understand each other, even if they both comply with the JDF specifications. The reason for this is that the developers of a system do not implement the entire JDF functionality, but consciously or unconsciously expect certain prerequisites. A JDF reading device may simply ignore certain parts of the JDF information provided. For example, it is conceivable that a folding machine reads the resource *FoldingParams* properly, but then only interprets the information about the *FoldCatalog*, not the individual folding operations. After all, both attributes are marked as optional in the JDF specification.

Of course, there are also new versions of JDF/XJDF published from time to time. Certain new features are introduced, and old ones are discontinued. This alone might lead to incompatibilities.

The *Interoperability Conformance Specification (ICS)* is intended to remedy this situation. These papers describe which parts of the specification should be observed for certain production sectors. There are different papers for different sectors and interfaces. For example, there is an ICS paper for MIS-to-Prepress, another one for MIS-to-Finishing, and so on.

The ICS papers also define names of Gray Boxes. For example, you will look in vain for information about Gray Box *PlateMaking* in the JDF specification. Instead, there is a description in *MIS to Prepress ICS Version 1.5* (CIP4 2015b).

The Gray Box name is noted in the *Category* attribute of the JDF node. The ICS papers define which process names must be entered in the *Types* attribute of the Gray Box (for example by the MIS) and which must be read (for example, by the prepress system). These two indications may be different, because sometimes a process may be written optionally but must be read if it is present. Moreover, certain conditions can be defined, for example, by saying that from two possible processes only one should be listed. The ICS papers also define for Gray Boxes which input and output resources must be written and read. Here again, conditions can be set. For example, the attribute *FoldCatalog* must be written in the *FoldingParams* resource if folding positions (*Folds*) are not specified.

It is part of the nature of a Gray Box that not all required process resources are defined. For instance, a *RunList* holding images (bitmaps) is a necessary input for the *ImageSetting* process. Furthermore, *ImageSetting* is a necessary process of the *PlateMaking* Gray Box. However, there is no statement in the MIS-To-Prepress ICS about such a *RunList* resource in the *PlateMaking* Gray Box. This resource should be generated later during the expansion of the Gray Box.

Many ICS papers do not describe a single conformance specification but rather up to three. They are called *levels*, where Level 3 includes Level 2, and analogously Level 2 includes Level 1. These three levels represent different compatibility requirements. In simple terms, the levels identify low, medium, and higher compatibility conditions. For example, according to *MIS ICS Version 1.5* (CIP4 2015c), an MIS only needs to read the *ResourceAudit* resource element in an *AuditPool* if it is compatible with Level 2, and not if it is compatible with Level 1. The *ResourceAudit* element describes the usage of resources during execution (see Section 4.3.4).

Both JDF and XJDF optionally provide the *ICSVersion* attribute to state which ICS specification they comply with. The attribute should be entered in the root JDF Node but may also be entered in others. The entry *ICSVersions="Base_L2-1.5 MIS_L2-1.5"* in Figure 4.26 states, for example, that the JDF node meets the conformance requirements of both Level 2 of the *Base ICS Version 1.5* (CIP4 2015d) and Level 2 of the *MIS ICS Version 1.5* (CIP4 2015c).

```
<JDF xmlns="http://www.CIP4.org/JDFSchema_1_1"
Version="1.7" Type="Product" Status="InProgress"
JobID="07-0111" ID="4711" ICSVersions=
"Base_L2-1.5 MIS_L2-1.5">
```

*Figure 4.26:
Example of the opening
tag of a JDF root element*

If two systems have compatibility problems, it might be a good idea to check if they comply with the same ICS levels.

4.7 Portable Document Format (PDF)

Since the earliest days of PDF, the format could contain a *Document Information Dictionary*. This dictionary can only record a few values such as the document's title, author's name, and creation/modification dates. A *Tagged* PDF, which was introduced 2001, contains additional metadata information for the document structure. This means that PDF is no longer just an amorphous page description language, but that the content can be structured in headings, paragraphs, and the like. This logical structure of the page content allows, for example, automatic reflow of text, PDF conversion to XML/HTML documents, and defining the reading order for text-to-speech. In 2004, Adobe published the *Extended Metadata Platform (XMP)* specification (see Section 4.1). It is mainly used to store information about page components (such as images) or about the PDF document itself (such as the author's contact details). Because XMP metadata is much more flexible than the *Document Information Dictionary*, the latter was discontinued with the PDF 2.0 specification (ISO, 2017).

Both PDF tags and XMP metadata are rarely used for specifying the product or controlling the production processes. The XMP structure is not suitable for defining properties of product parts. We covered this in Section 4.1. Tagged PDF structures the layout of the content, but not in accordance with the requirements of manufacturing processes.

This changed in 2010 with the introduction of PDF/VT. This format was designed for document exchange concerning variable data and transactional (VT) printing. The idea was to reduce the amount of processing time for printing variable data. This is achieved by creating another branch for document parts (*DPart* tree) in parallel to the PDF page tree. This allows document parts to be identified within a PDF file by setting page starts and page ends referring to the classic page tree of the PDF file (see Figure 4.27). Each *DPart* object can also be associated with its own *Dictionary Document Part Metadata (DPM)*, which in turn can hold application-specific information about the document parts. The standard defines only the DPM structure that can be populated with basically anything; it does not define specific entries to be used within that structure. For a VT application, it may hold the address of every recipient, for instance. The PDF/VT file must comply with PDF/X-4 or PDF/X-5 specifications.

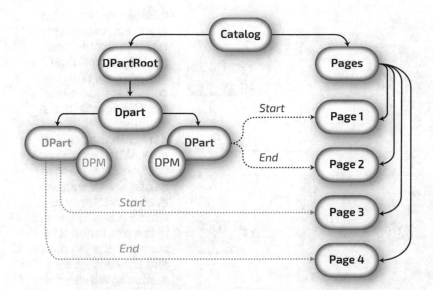

Figure 4.27:
The DPart tree specifies
sub-documents (parts) of
a PDF file..

4.7.1 Print Product Metadata

The product intent of a print buyer is normally described in JDF and XJDF. That is, both formats can hold properties of the requested product, such as the product type, printing substrate, binding intent, type of inks to be used, number and dimensions of pages, and delivery details. In JDF, the *Intent Resources* define the details of products to be produced, In XJDF, the *Product Intents* take over this role.

Since 2019, there has been third way to store this kind of data – inside PDF. The specification for this is in (ISO 2019b). It is based on (CIP4 2010). See also (Hoffmann-Walbeck, 2018).

Thus, a PDF file contains not only the artwork data but also the product description. This concept goes far beyond considering PDF as just a page description language (PDL).

The idea is that the print buyer stores the product intent inside the PDF file, either by entering the data directly in some layout program, or indirectly by using, for instance, a web-to-print systems that writes this metadata automatically into the PDF file. When such a PDF file reaches the print provider, it can be split up into two parts: PDF artwork data and JDF/XJDF product intent data. As a result, the print shop does not need to change its workflow automation. On the other hand, the advantage of such an approach is that both artwork data and product intent data are stored in one single file during transmission between print buyer and print provider. There is no need any more to link the artwork and the job ticket at the print provider's site. Figure 4.28 illustrates this scenario.

The difference between the classical and the new concept in Figure

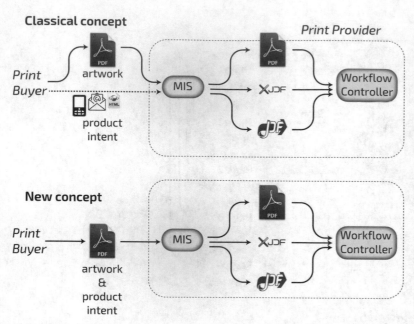

Classical concept

Print Buyer

artwork

product intent

Print Provider

MIS

Workflow Controller

New concept

Print Buyer

artwork & product intent

MIS

Workflow Controller

Figure 4.28: The Product Intent is transported via PDF

4.28 seems very subtle, but the new concept might boost the print provider's workflow automation. With the current standard production approach, the PP must link the job data with the artwork data. With the new concept, this is no longer necessary. Of course, even now it is not always a human operator in the print shop who connects the job data and the content data manually, although this is still common practice. By using a portal or a W2P system, this can also be automated. After all, it is still common for the PB to send the job data via e-mail and the content data via a file transfer service (see Section 2.2.1). In this context, the new concept might become useful.

Please note that Figure 4.28 does not show all possible conversions of the PDF file to a job ticket format at the PP's site. In particular, the data format CSV should be added that some prepress systems can process.

How is the product description data stored inside a PDF? It uses the *DPart* structure as shown in Figure 4.27 for PDF/VT. This mechanism allows to logically split a PDF file into several chunks, for instance into pages for the cover and others for the content. The product intents can then be stored in the associated *Document Part Metadata (DPM)* using standard keywords. A *DPM* is not coded in XML such as JDF and XJDF, but rather in a PDF dictionary. Figure 4.29 shows a snippet of *DPM* code. A dictionary in PDF is a table containing one or several key/value pair(s) – like a normal dictionary for languages. Each dictionary is wrapped by double angle brackets, that is <<key value key value...>>. Unlike a normal language dictionary, the value of a PDF dictionary can be an (almost) arbitrary object, for example another dictionary. That is why the structure in Figure 4.29 looks a bit wild. Each dictionary represents a level in the hierarchy for DPM structure. After all, the keywords are easy to recognize: they all start with the letters */CIP4_*.

Finally, I would like to emphasize that the job submission approach presented here is a possible concept for the future, but probably has not yet been implemented on a large scale so far.

```
<</DPM
  <</CIP4_Root
    <</CIP4_Production
      <</CIP4_DescriptiveName(Cover)
      >>
      /CIP4_Intent
      <</CIP4_ProductType(WrapAroundCover)
        /CIP4_MediaIntent
         <</CIP4_MediaQuality(lumisilk_135)
           /CIP4_MediaWeight(135)
         >>
      >>
    >>
  >>
>>
```

Figure 4.29:
Code snippet of a DCM
structure

4.7.2   PDF Graphic Objects as Processing Steps

It is common to use graphical objects in PDF as information for production. Examples can be found especially in the packaging sector, such as die cutting, gluing, creasing, braille, etc. PDF paths are not used for drawing but rather for controlling some production processes. Normally, the data producer creates separate layers and/or special colors for this purpose. This works fine, but there is a catch: the names of the layers and colors are not standardized. Thus, the data creators must always send a separate note to the data consumer in order to explain the PDF. Furthermore, an automatic preflight program cannot check the PDF according to these special requirements. For example, if a color defines a structural design, it must be a spot color, and the object must be set on *overprint*. However, it can easily happen during PDF generation that a spot color is converted to CMYK or that layers are merged. Sometimes people simply forget to set the graphical object to the overprint mode. These kinds of data are then no longer useful.

The ISO 21812-1 (ISO 2019) standard intends to improve this situation. It defines how a graphic can be assigned to a process step. For this purpose, a so-called *Optional Content Group* (OCG) is created which contains a *GTS_Metadata Dictionary* (see Figure 4.30). The process step is defined in this *GTS_Metadata Dictionary*. In the example shown, this is *Cutting*. In addition, there are 22 other values defined in the ISO standard, such as creasing, drilling, gluing, foil stamping, embossing, folding, etc. It is up to the PDF processor to use this information appropriately.

Let us conclude this section with a brief summary of the PDF metadata that we discussed here:

```
7 0 obj
  <<
    /Name (Die cut)
    /GTS_Metadata <<
      /GTS_ProcStepsGroup /Structural
      /GTS_ProcStepsType /Cutting
    >>
  >>
endobj
```

Figure 4.30:
The PDF snippet shows
an OCG object with a
GTS_Metadata key in the
OCG dictionary

- *Document Information Dictionary* for storing a small set of fixed keywords.
- *XMP* defines metadata on the page and page element level.
- *Tags* define an HTML or XML kind of structure for PDF, such as header, section, paragraph, figure, etc.
- *Object Content Groups Metadata* for packaging and labeling.
- *PDF/VT* makes it possible to split PDF documents for different recipients.
- *DPart* Product Intent allows product descriptions to be defined in PDF. The print buyer can store suitable metadata in PDF, which is then used to convert it into a job ticket at the print provider's site.

References

Adobe Systems Incorporated (2012), XMP Specification Part 1, Data Model, Serialization, and Core Properties. Available at: https://ww-wimages2.adobe.com/content/dam/acom/en/devnet/xmp/pdfs/XMP%20SDK%20Release%20cc-2016-08/XMPSpecificationPart1.pdf (Accessed: 15 June 2021).

Adobe Systems Incorporated (2016), XMP Specification Part 2, Additional Properties. Available at: https://www.adobe.com/content/dam/acom/en/devnet/xmp/pdfs/XMPSDKReleasecc-2020/XMPSpecificationPart2.pdf (Accessed: 15 June 2021).

Adobe Systems Incorporated (2020), XMP Specification Part 3, Storage in Files. https://www.adobe.com/content/dam/acom/en/devnet/xmp/pdfs/XMPSDKReleasecc-2020/XMPSpecificationPart3.pdf (Accessed: 15 June 2021).

CIP4 2010: ICS — Common Metadata for Document Production Workflows. Available at: https://confluence.cip4.org/display/PUB/Common+Metadata+for+Document+Production+Workflow+ICS (Accessed: 15 June 2021).

CIP4 2015a: PrintTalk Specification 1.5 (2015). Available at: https://confluence.cip4.org/display/PUB/PrintTalk (Accessed: 15 June 2021).

CIP4 2015b: MIS to Prepress ICS Version 1.5 (2015). Available at: https://confluence.cip4.org/display/PUB/MIS+to+PrePress+ICS (Accessed: 15 June 2021).

CIP4 2015c: MIS ICS Version 1.5 (2015). Available at: https://confluence.cip4.org/display/PUB/MIS+ICS (Accessed: 15 June 2021).

CIP4 2015d: Base ICS Version 1.5 (2015). Available at: https://confluence.cip4.org/display/PUB/Base+ICS (Accessed: 15 June 2021).

CIP4 2018: XJDF Specification 2.0 (2018). Available at: https://confluence.cip4.org/display/PUB/XJDF+2.0 (Accessed: 15 June 2021).

CIP4 2020a: JDF Specification 1.7 (2020). Available at: https://confluence.cip4.org/display/PUB/JDF (Accessed: 24 June 2021).

CIP4 2020b: XJDF Specification 2.1 (2020). Available at: https://confluence.cip4.org/display/PUB/XJDF (Accessed: 15 June 2021)

CIP4 2020c: PrintTalk Specification 2.1 (2020). Available at: https://confluence.cip4.org/display/PUB/PrintTalk (Accessed: 15 June 2021).

CIPA (2016), CIPA DC- 008-Translation- 2016: Exchangeable image file format for digital still cameras: Exif Version 2.31, Camera & Imaging Products Association, https://www.cipa.jp/std/documents/e/CIPA_DC-X008-Translation-2016-E.pdf (Accessed: 30 June 2021).

Consignor. Available at: https://www.consignor.com/carriers/ (Accessed: 15 June 2021).

DIN 66001 (1966), Deutsche Institut für Normung e.V., Sinnbilder für Datenfluß- und Programmablaufpläne, Beuth Verlag GmbH. Available at https://standards.globalspec.com/std/639497/DIN%2066001 (Accessed: 15 June 2021).

Hoffmann-Walbeck T (2018): PDF Metadata and Its Conversion to XJDF, GRID 2018 Proceeding, p. 445-453. Available at: https://www.grid.uns.ac.rs/symposium/download/2018/grid_18_p54.pdf (Accessed: 15 June 2021).

ISO 2010: ISO 16612- 2: 2010, Graphic technology — Variable data exchange — Part 2: Using PDF/ X- 4 and PDF/ X- 5 (PDF/ VT- 1 and PDF/ VT- 2). Available at: https://www.iso.org/standard/46428.html (Accessed: 15 June 2021).

ISO 2017: ISO 32000-2, 2017, Document Management - Portable Document Format - Part 2: PDF 2.0. Available at: https://www.iso.org/standard/63534.html (Accessed: 15 June 2021).

ISO 2018: ISO 19593-1:2018, Graphic technology — Use of PDF to associate processing steps and content data — Part 1: Processing steps for packaging and labels. Available at: https://infostore. saiglobal.com/en-us/Standards/ISO-19593-1-2018-1128856_SAIG_ ISO_ISO_2618548/ (Accessed: 15 June 2021).

ISO 2019a: ISO 16684-1:2019, Graphic technology — Extensible metadata platform (XMP) — Part 1: Data model, serialization and core properties. Available at: https://webstore.ansi.org/Standards/ ISO/ISO166842019 (Accessed: 15 June 2021).

ISO 2019b: ISO 21812-1:2019 Graphic technology — Print product metadata for PDF files — Part 1: Architecture and core requirements for metadata. Available at: https://www.iso.org/standard/74407.html (Accessed: 15 June 2021).

Meissner S., XJDF - Exchange Job Definition Format, ISBN 978-3-00-055604-3 (2017). Published at ricebean.net.

Meissner S., EasyXJDF (2018), available at: https://confluence.cip4.org/display/PUB/EasyXJDF?preview=%2F688397%2F29426299%2FCIP4+EasyXJDF+Bologna-18.03-bin.tar.bz2 (Accessed: 26 June 2021).

5 Glossary

Agent	An agent is a workflow component that initiates the workflow communication by writing the initial job ticket for a print job.
B2C	Business-to-customer (B2C) denotes a business interaction between an end consumer and a company.
B2B	Business-to-business (B2B) refers to a business relationship between two (or more) companies.
Business Object	Business objects or business data describes a business transaction between print buyer and print provider, such as a request for a quote, a purchase order, or an invoice.
Byte Map	Data structure for images with color depth greater than one (e.g., eight) per separation.
CFF2	CFF2 stands for Common File Format Version 2. It is a text format for CAD data.
CIP3	CIP3 stands for International Cooperation for Integration of Prepress, Press and Postpress. It is the predecessor organization of CIP4.
CIP4	The International Cooperation for the Integration of Processes in Prepress, Press and Postpress is a non-profit standards association organization specifying data formats for workflow automation in the printing industry.
Cobots	Cobots, or collaborative robots, are designed to share a workspace with humans.
Component	A JDF Component describes the various versions of (semi-) finished goods. The final product, product parts, but also a set of printed sheets or a pile of folded sheets are considered components.
Controller	A controller routes job tickets and messages in workflow system(s) and devices(s).
CSV	CSV stands for **c**omma-**s**eparated **v**alues and denotes the structure of a text file for storing or exchanging spreadsheet data.
CtP	Computer-to-Plate (CtP) is a standard procedure to image the print data directly on a printing form.
DAM	DAM stands for digital asset management. It allows the storage, retrieving, organizing and manipulation of digital assets such as images.
Device	A device is a component of a workflow part that interprets a job ticket and initiates the execution of one or several processes.

DTP Point	A Desktop Publishing Point equals 1/72 Inch. That is about 0,36 mm.
DUNS	DUNS stands for Data Universal Numbering System. It is a unique nine-digit identifier for businesses.
EDI	Electronic Data Interchange (EDI) is a universal term for the electronic transfer of information from one computer system to another.
Gang form	In a gang form (parts of) several print jobs possibly from different customers are placed on the same sheet.
ICS	ICS is a set of specifications developed by CIP4. ICS documents specify the interface requirements how JDF/XJDF modules (e.g. MIS, Prepress controllers or Press controllers and devices) should interoperate if they comply to the same ICS level.
Imposition scheme	An imposition scheme (often called folding scheme) indicates how pages are placed on a sheet. Only the principal placement of the pages in relation to each other is defined, not the exact position.
Interface	An interface is the point where two processes or software components interact.
Interpreting	Interpreting is the first process inside a RIP. PDL data like PDF is read, parsed and simplified for the next process. For example, Bézier curves are iteratively approximated by polygons, images are decompress etc.
Job ticket	A job ticket is a document that details the specifics of a print order. It can contain a description of the intended print products as well as instructions for their production.
JSON	JavaScript Object Notation (JSON) is a text-based, structured data format. It is similar to XML.
JMF	JDF goes together with the Job Messaging Format (JMF). This allows sending feedback back from devices to controlling software. The protocol between the parties supports dynamic, real-time interaction, e.g. a controller can query a device for job tracking or send commands concerning the device's input queue.
Metadata	Strictly speaking, metadata describes properties of other data. For example, if the data represents a document that includes text and formatting rules, the corresponding metadata might contain properties such as the author's name, a copyright notice, and the creation date.
Message	Messages enable data exchange in near real time. Messages for the graphical industry are defined in JMF and XJMF.
Namespace	A namespace and an associated prefix is specified by an xmlns

attribute. XML elements can then be placed in any namespace by a prefix, which is separated from the element name by a colon.

Online printer

An online printer is a print provider that gets orders (almost) exclusively via one or several W2P shops.

PDF Metadata

PDF metadata in this context refers to a standard from 2019, which allows to write product intent data into a PDF file. The information can be, for example, about the require printing substrate, the binding intend or even which pages of the PDF file are defining the cover and which the content.

OEE

The overall equipment effectiveness measures the manufacturing performance relative to its full potential, i.e., the manufacturing productivity.

PDL

PDL stands for Page Description Language. PDF is the best-known PDL.

PJTF

PJTF stands for Portable Job Ticket Format. Adobe published the specification in 1999.

PPF

The Print Production Format (also called CIP3 Format) is a predecessor of JDF, which is still widely used in ink key presetting systems. The Print Production Format has been specified by CIP3, from which CIP4 emerged. Today, CIP4 maintains PPF.

PrintTalk

PrintTalk is an open XML standard from CIP4 used to communicate business information. It provides JDF and XJDF with capabilities for pricing, web-to-print, RFQ/quote, invoicing, order status, sub-contracting and more.

Private XML

Private XML is vendor specific XML.

Process

A process is a singular activity with a specific objective that can be planned and executed independently.

Product Part

A product part is a component of a product. For example, a softcover book consists of two parts: cover and content.

Product Specification

The product specification is a description of how the print buyer wants the product to be in the end. It is independent of the necessary production processes.

Resource

A resource is either a physical or an electronic/conceptual object.

Regular Expression

A regular expression is a sequence of characters that defines a pattern.

Rendering

Rendering converts graphical elements, which might include vector data, into contone images (byte maps).

Responsive Web Design

With responsive web design, the function, design and content of a web page might change in accordance with the screen size and resolution of the computer monitor, tablet, smartphone, etc. being used.

RIP A Raster Image Processer renders PDF data (or some other PDL) into images.

RunList A RunList defines one or more printable logical documents, e.g., a PDF file.

Screening The screening process converts contone images to monochrome images (bitmaps).

Transfer Curve A transfer curve is a curve/table that defines modifications of tonal values of the artwork during RIPing to match desired tonal value increases in print – for example, to reach an offset standard.

VPC A virtual private cloud (VPC) is a private cloud within a public cloud environment.

Web Service A web service is a software system designed to support interoperable machine-to-machine interaction over a network.

Web-to-Print (W2P) W2P is a commercial printing platform for job submissions using web sites.

WMS A workflow management system (WMS) is a software system that defines, creates, and manages the execution of workflows. It also controls one or more devices which are able to execute processes.

Workflow A workflow denotes the sequence of production and business processes, in whole or part, during which documents, information and task descriptions are passed from one participant to another for action according to a set of procedural rules.

Workflow Logic The workflow logic is the description of rules and relationships between processes and resources that are needed for print production.

XJDF XJDF is an interchange format between two applications, for example, between a controller and devices.

XML XML stands for Extensible Markup Language. It is a very popular way to structure any data in text format. Standard data formats like JDF and XJDF are XML based.

XML Schema An XML schema is a description of the structure of an XML document type. In particular, a schema defines the names of elements and attributes, the hierarchy of elements, and the data types of attribute values.

XSL XSL stands for Extensible Stylesheet Language. It is used, for example, to transform an XML format into another XML format.

6 Index

Printed in the United States
by Baker & Taylor Publisher Services